FUELLING THE MOTORING AGE

FUELLING THE MOTORING AGE

100 YEARS OF BRITISH PETROL STATIONS

NICK EVANS

The History Press

Front cover image: Turnbull's circular self service petrol station in Plymouth, Devon, which opened in April 1963, featured an unusual cantilever frame to obviate the need for pillars and keep the floor space free of obstruction. The place has long since been demolished. (Photographic Services Shell International)

Back cover images. Top to bottom: The Old Store at Nether Lochaber, Onich, Angus (Chris Barker collection); Restored 2-gallon petrol cans at the Old Airport Garage Museum, Rochester, Kent (Author's collection); BP Chargemaster (BP Chargemaster/Polar)

First published 2019

The History Press
97 St George's Place,
Cheltenham,
GL50 3QB
www.thehistorypress.co.uk

British Library Cataloguing in Publication Data.
A catalogue record for this book is available from the British Library.

ISBN 978 0 7509 9149 0

Typesetting and origination by The History Press
Printed in Turkey by Imak

CONTENTS

INTRODUCTION

Petrol stations have now been with us for 100 years. Their existence, while vital to fuelling our vehicles, as well as powering the national economy, has largely been unheralded throughout that time.

Indeed, the idea of a purpose-built petrol station wasn't devised until cars had been running on British roads for more than a quarter of a century – and it took a motoring organisation rather than an oil company to show us all what to do back in November 1919.

Whatever car, van or lorry you drive, you have to visit a forecourt sooner or later. For the many millions who spend any time at the wheel, pulling into a petrol station is just a necessary chore – something to be endured rather than enjoyed. Having filled up we hand over a chunk of money, wanting to get the experience over and done with as quickly as possible.

Today's petrol stations tend to be bland, instantly forgettable places – essentially a large canopy covering rows of multi-hosed box-shaped pumps with a cashier inside a convenience-style shop a short distance away. Those forecourts of yesteryear, while not normally considered glamorous, had character and individuality, whether they were set in a rural backwater, at a kerbside or formed part of a bustling townscape that reflected the latest architectural style. An attendant would emerge from a kiosk to fill up your car for you, wipe the windscreen and – if you asked nicely – check the oil and tyres too.

Opposite: Set on a corner of the Epsom Road in East Horsley – a small village part way between Guildford and Leatherhead in Surrey – the pump attendant sets to work filling up an early Ford Anglia with BP Super, priced 4/8*d* (23p) per gallon. The turreted building behind is the former twin-towered gatehouse to Horsley Towers. By 2007 at least, BP pumps had given way to Murco. (© BP plc)

Like other components of the retail scene, the number of petrol stations has been diminishing for many years – the brunt borne largely by the independents tired of holding out against bureaucracy, slim profits and blatant theft. Just as the petrol station concept reaches its centenary milestone, it may not be very long before there are even fewer forecourts to drive to as we adopt cleaner electric and alternative-fuel vehicles.

Meanwhile, enjoy this book as a nostalgic reminder of British petrol stations at their best and how they have played their part in keeping our wheels turning. While many of the examples included may no longer exist, redevelopment into housing is common and some live on to fulfil another need.

I'm particularly indebted to Chris Barker and Ed Coldrick for letting me have unfettered access to their terrific collections of petrol station memorabilia. Also to Trevor Hoare, who owns the Old Airport Garage Museum at Rochester Airport in Kent, who kindly allowed me to photograph many of the exhibits there and to the thousands of followers of the Facebook page The Golden Age of the Great British Petrol Station who inspired this book in the first place.

Nick Evans
Whitstable, Kent

FOREWORD

I share my passion for Britain's characteristic roadscape with Nick Evans, but I hadn't quite realised that the country's petrol stations are only just reaching their centenary. As Nick says, visiting one today is a chore for most of us, but the earliest drivers had to wait thirty-five years for them to arrive. In the interim, they bought petrol at chemists' shops and carried it around in tin cans. The inconvenience – not to mention the combustible peril – seems utterly unimaginable today.

I first met Nick at a conference staged to help encourage new motoring authors, where I was speaking on the joys (or otherwise) of research. It's obvious to me that he went away brimming with determination to communicate a little-chronicled area of British car culture.

As a consequence, he's poured a huge number of facts, statistics and anecdotes – and plenty of perspectives and context – into this look at the British petrol station. I feel it helps explain much about our motoring culture that would be difficult to appreciate were it not for the way he's 'curated' it all.

It's a complex journey – with, of course, regular stop-offs – but it's also very timely, because a dramatic shift to electric and even hydrogen power looks likely to reshape the petrol station one final time. Before they vanish, Nick has done a wonderful job in explaining their evolution.

Giles Chapman
Motoring author and magazine editor
April 2019

A gleaming Ford Anglia is filled by a slightly stony-faced attendant at Wilsons Motors in Spalding, Lincolnshire, during the early 1960s. Cleveland's least-expensive petrol is on sale at all but 5 shillings (25p) per gallon. Behind the pumps, a new Hillman Imp – or perhaps it's the more upmarket Singer Chamois – is for sale. (Ed Coldrick collection)

1 FUELLING UP — THE FIRST SPARKS

Petrol is not a new substance; its origins can be traced to seventh-century Persia, China and Japan where, known as 'burning water', it had a variety of medicinal purposes.

Petrol, as we recognise it, was patented by Derbyshire mine owner James Young in 1850 two years after he discovered the spirit leaking into his Riddings colliery, near Alfreton. Initially, after he distilled it, the fluid he extracted was used for lighting lamps and for lubricating machinery. Later, he also produced paraffin after further developing his distillation process.

The first cars took to the roads in the mid-1880s, but it would be at least thirty years before the first purpose-built petrol station opened for business in Britain.

During that time, some considerable motoring milestones were passed – doing away with a man walking in front of your car bearing a red flag to warn others of your approach in 1896 was the most liberating one.

Motor, or petroleum, spirit simply became known as petrol when the Carless, Capel and Leonard Company in 1899, the leading refiners of the day, registered it as a trade name. It had been suggested by Frederick Simms, a co-founder of the Royal Automobile Cub and co-patentee (with Robert Bosch) of the magneto.

Simms was importer of the first Daimler into Britain and was hoping to allay people's fears about storing flammable petroleum spirit. His logic was that the term 'petrol' didn't sound particularly dangerous and people wouldn't worry too much about it. The name has stuck ever since.

Petrol was supplied in specially designed 2-gallon cans for the first time from around 1898 by Pratt & Co., and this would be the principal means of sale until petrol pumps became widespread in the 1920s.

Britain's first purpose-built petrol station was opened by the Automobile Association in November 1919 in a layby on the A4 Bath Road, 2 miles from the Berkshire village of Aldermaston. The AA opened seven more petrol stations in the early 1920s, keen to show their practicality rather than make a profit. As the idea caught on, so the AA closed or sold them, the last going by 1932. Any trace of this first petrol station has long been lost.
(Author's collection)

Head out of London on the Great West Road today and you will meet miles of traffic! The Ace Motor Company, seen here in 1928, had a fine line of Shell pumps to serve passing customers. Clearly, trade was slow when the picture was taken – just one car and a motorbike with sidecar are using the A4. A sturdy direction sign denotes Southall is 3 miles away on the B358 (now called the A3063) while Teddington and Kingston are both 4 miles distant. (Chris Barker collection)

Anderson's Garage on the Ayr Road at Newton Mearns, a few miles south of Glasgow, originally started business as a cycle agent but by the 1920s had become a fully fledged motor garage serving Pratt's and Shell petrols. Was it by chance that tankers representing both brands were replenishing the underground tanks when the photo was taken or was it carefully pre-arranged? We may never know. Anderson's was a successful Rootes Group main dealer for many years in the twentieth century.
(Author's collection)

The Automobile Association (AA) and the Royal Automobile Club (RAC) had been formed and chalked up notable successes in improving the lot of the early motorist by pursuing changes in the law, particularly relating to roadside speed traps.

Known to residents of Beckenham, Kent, as the Chinese Garage, it was designed as a Japanese pagoda in 1928 by Edmund Clarke. In 1930, the Chinese Garage – officially Langley Park Garage – won first prize in a Better Petrol Stations competition run by the *Daily Express* and the Gardeners' Guild. It became a grade-two listed building in 1989. In August 2017, the local paper revealed the site, which had long ago ceased to sell petrol, might become a Tesco Express outlet. (© BP plc)

Asking motorists not to smoke while their cars are being filled up: the driver here has his pipe out ready for a sneaky puff while the attendant is distracted by operating the National Benzole pump at an unknown location sometime in the 1920s. As the motorist is holding the pump nozzle to his tank, could this be considered an early form of self service? (© BP plc)

Many small and large manufacturers were finding their stride in building cars – Rolls-Royce was formed in 1906, launching its legendary Silver Ghost model soon after. Cars moved from being built in sheds to purpose-built factories, especially once Ford pioneered the moving assembly line in the years immediately before the First World War.

Motor racing had become a popular sport for the wealthy with speed and endurance records being set and broken at regular intervals.

Perhaps unsurprisingly, the government introduced petrol tax in 1909 at the rate of 3*d* (1.5p) per gallon. This was abolished by 1920 but made a permanent comeback in 1929.

The first roadside fuel pump in Britain was erected and connected to a 2,000-gallon underground tank in Shrewsbury as early as 1912 – although it may have been as late as 1915 depending upon who you want to believe – at Legge & Chamier's garage in Abbey Foregate. Sited at the kerb and cranked by hand, the pump was made by Bowser of Fort Wayne, Indiana, USA.

The garage later became the West Midland Motor Company and, by 1962, the Bowser pump had found its way to the Weights & Measures department of the local council. Some years after, the pump was purchased by an anonymous collector after he discovered it languishing in a nearby builder's yard.

The kerbside pump in Abbey Foregate, Shrewsbury, thought to have been the first in Britain, appears to be serving National Benzole when this picture was taken, perhaps after the First World War. (*Shropshire Star*)

Bentalls is a long-established classy department store in Kingston-upon-Thames in Surrey and its affluent customers would have been pleased to see it was keeping up with modern times when it opened its own petrol station during the 1930s. The move was most likely to have been prompted by expanding housing development and car ownership in the London suburbs. The overall canopy suggests some forward-thinking design in an era when many forecourts were still open to the elements. (Ed Coldrick collection)

At first, he told local newspaper the *Shropshire Star* he didn't realise its significance until he started researching the pump's backstory. He told the paper: 'I don't know if it works, although all the mechanics do function. I don't put any petrol through it. Regulations are so strict you wouldn't want to. It last worked 40 or 50 years ago.'

The Great War of 1914–18 intervened but proved the worth of motorised vehicles, as well as the need to protect petrol supplies, in moving men and materials around the Western Front. A fleet of London buses was requisitioned by the War Office as troop transport in parts of northern France, while motor ambulances returned countless wounded to field hospitals and lines of lorries moved supplies to where they were needed.

With the war over, motorists were content to continue buying their petrol in metal 2-gallon cans from a variety of chemists, hardware shops or indeed the many blacksmiths that had transformed themselves into garages. Often, these cans were mounted on car running boards and used as needed.

During a national rail strike in 1919, the AA set up temporary petrol dumps in some city car parks enabling members to replenish their tanks more easily while using their cars in place of trains. That November the AA took it upon itself to build a wooden hut with canopy beside the Bath Road, 2 miles from the Berkshire village of Aldermaston – effectively creating Britain's first purpose-built petrol station.

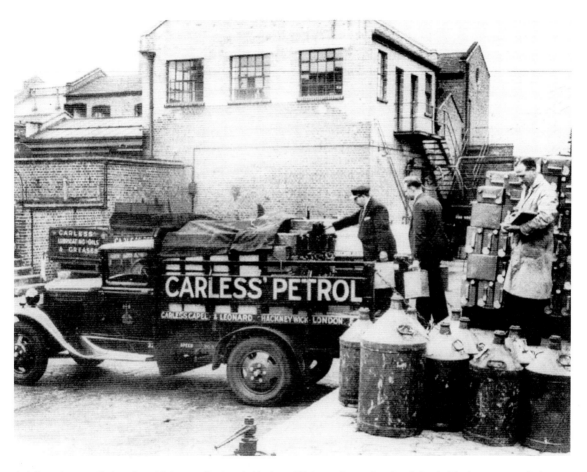

A delivery lorry at Carless Capel & Leonard's plant in Hackney Wick, north-east London, is loaded for its next round of deliveries sometime *c.* 1919. The company christened motor spirit 'petrol' as it sounded less dangerous to an unsuspecting public. Note the stack of carefully arranged cans behind the clerk on the loading platform. (Author's collection)

Inside the hut was a large drum of petrol with a hand pump attached. A pipe would be fed through the front window by a uniformed AA patrolman to the member's car. In those days filler caps were usually dashboard-mounted so considerable care was needed to avoid spillages near a hot engine!

The venture also enabled the AA to promote the sale of British-made benzole – a by-product of burning coal – in preference to importing Bolshevik Russian petrol.

The AA went on to open several more filling stations around the country during the early 1920s, keen to show their practicality rather than to make a profit. As the idea caught on, so the AA closed or sold them, the last going by 1932 by which time there were nearly 60,000 places selling petrol across Britain. A comprehensive supply network had been established.

These roadside refuelling facilities ranged from the ramshackle to those incorporating the trendiest building design features of the period.

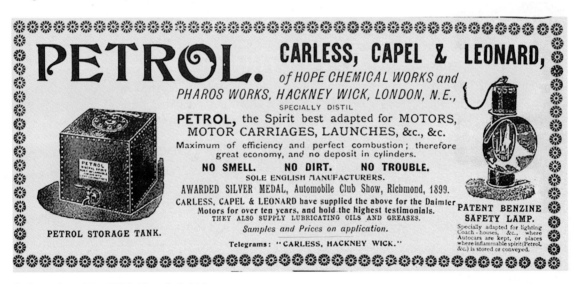

Carless advert, May 1904. (Grace's Guide)

Looking in good order just after opening during the late 1920s is Burbridges petrol station on Humberstone Road, Leicester, which has been adorned with advertising hoardings around its semicircular perimeter to be seen by customers. Inward-bound cars are directed to go to the left of the pumps – hopefully filler caps were mainly on the offside of the vehicle! An attendant waits patiently in the doorway of his kiosk for some trade. Today, the site is a canopied Texaco site with an off-licence attached. (Chris Barker Collection)

We can date this photo of the village of Cushendell in County Antrim, Northern Ireland, to the mid- or late 1930s as Esso's Essolene pumps are clearly on view to the right here, outside the sweet shop and tobacconist. Across the road, a solitary Shell pump stands tall behind a low wall where a small group has gathered, including local police, to provide a focal point for the photographer. (Author's collection)

In more rural areas there might be a single isolated pump with a small corrugated iron hut offering some cover for the attendant. The pump would be connected to a barrel, the hose being held in place by a cloth bung – in enclosed spaces, this also served to reduce the escape of petrol vapours.

In towns, the pump might be right by the kerb outside an operator's premises. Sometimes, there could be two or three outside a shop, some yards apart and offering different brands and grades of petrol. By now many motorists had found a particular brand might better suit their car's engine and tended to stick with it whenever they could.

Where pumps had to be set back against a shop front, either for easier bulk storage below ground or because the pavement was too narrow, so overhead swinging arms would be fitted to the pumps and pulled out across the pavement when serving to avoid the hose being trailed along the ground.

Cars gradually became more affordable through the 1920s as manufacturers slashed their costs and more people took up driving, licences having been required since April 1903. In turn, there was growing demand for petrol. Chemists and hardware shops became wary of selling larger quantities of 2-gallon cans, worried about the fire risks of storing them, and so were usually happy to see roadside pumps take their place. By the end of the 1920s, sales of petrol in cans were fast declining.

Top row, left: A rare surviving example of the earliest, and simplest, form of petrol pump is this 1904-origin metal barrel with a crank handle and rubber pipe affixed to its top. It was originally supplied by Anglo-American Oil, which would later become part of the company that evolved into Esso.

Top row, centre: An early example of a hand-operated petrol pump, now restored. Note the round-topped shade at the top with light bulb under, suggesting this preceded the idea of the popular glass globe. Beside it are some examples of 2-gallon petrol cans.

Top row, right: A display of restored 2-gallon petrol cans.

All of these are on display at the Old Airport Garage Museum, Rochester, Kent. (Author's collection)

Bottom row, left: Example of 1930s early petrol pump dials. (Author's collection)

Bottom row, centre: The simple mechanics of a hand-cranked mechanical petrol pump dating from the 1920s -- part of the display at the Old Airport Garage Museum. (Author's collection)

Bottom row, right: A contemporary photo of what is believed to have been the first petrol pump used in Britain. Dating from 1912 and made by Bowser in America, it now belongs to an anonymous collector in Shropshire. (*Shropshire Star*)

Shops would be converted and often offer the services of a mechanic as well to become a two-pump garage. These premises wouldn't usually have a forecourt in built-up areas, but away from towns, where land was cheaper and more plentiful, there was a better chance of creating an off-road Tarmacked apron area upon which to set pumps and a cabin for the attendant.

Larger car dealerships, which already had workshops and employed mechanics to service customers' vehicles, also installed pumps on their premises.

The oil companies were keen to ensure their products were the most visible and handed out large numbers of enamel signs to petrol stations across the country. They did this to such an extent that conservationists pleaded with government to intervene when signs were thought to be overwhelming the surroundings, especially if the buildings they were fixed to weren't attractive either. The result was a toning down of the advertising and a move to more elegant styling for petrol stations, some following the art deco trends of the late 1920s and early 1930s.

Some of the more striking designs of the era saw frequent use of long, curved plate-glass windows in the frontages. In rural areas, others were made to look like barns to blend in with their surroundings – although the thatched roofs they sometimes had would be considered an unwise choice in today's safety-conscious climate. One petrol station, at Beckenham in Kent – and still just about standing in recent times – was built to resemble a Japanese pagoda. Its staff were expected to wear traditional Oriental costume to complete the effect!

The petrol station concept had reached the Shetland Islands off Scotland by 1922 when this BP roadside pump was the first to be introduced there at Bixter, about 10 miles north-west of Lerwick. (© BP plc)

The driver and passengers of this car keep a safe distance from the petrol pump and attendant while filling up outside Legge & Chamier's premises in Abbey Foregate, Shrewsbury. The pump is thought to have been the first in the country when fitted around 1912. (Ed Coldrick collection)

By now, there were four large oil companies supplying petrol across Britain. The largest, with more than 50 per cent of the market, was Anglo-American, with one of its brands being Pratt's. In 1935, the company would change its name to Esso. The other key players were British Petroleum and Shell-Mex – the two companies combining their distribution networks in 1932 – along with National Benzole and Redline-Glico.

Additionally, there were around a further dozen independent suppliers. They bought oil from the spot market via brokers and sold to local retailers. The two biggest independent firms were Power Petroleum, established in 1923, and Russian Oil Products, formed in 1925. Both companies distributed Russian petrol below the larger oil suppliers' prices.

Opposite: Young women refill 2-gallon cans of BP Motor Spirit in the early 1920s at Carless, Capel & Leonard's depot in Hackney Wick, east London. (Author's collection)

Empty 2-gallon cans of petrol are unloaded from a BP lorry at Carless, Capel & Leonard's depot in Hackney Wick, east London, in 1923 after collection from shops and garages. (Author's collection)

Listen to traffic bulletins on the radio and you may hear Henlys Corner mentioned. This is a junction in north London where the A1 and the North Circular Road meet before heading to the centre of the capital. The hotspot takes its name from the art deco-influenced Henlys garage, which stood between 1935 and 1989. The company had branches across the country claiming to be 'England's leading motor agents' and this branch was its flagship. (Mike Jeavons)

As the 1930s arrived, so greater motoring regulation was introduced – largely to improve public safety and reduce the rising number of fatalities on Britain's roads. By 1931, the first Highway Code had been published setting out the vital rules for safer motoring, and in 1935, people who wanted a driving licence had to first pass a test. Until then, they had only to apply at a Post Office for a licence, which was renewed every year.

By now, the big four were working with the Motor Trade Association and the Motor Agents' Association and had created a distribution network that would supply only to retailers and commercial consumers – and not direct to the public. Retailers had to be genuinely engaged in the motor trade on a site approved by the MAA.

At a stroke, pubs, cafes and other establishments that had been offering petrol as a useful sideline were excluded. However, many carried on by purchasing from the smaller suppliers. None of the retailers were tied to one single company and could continue to offer a selection of brands to their customers.

By the early 1960s it was still possible to come across the occasional kerbside petrol pump. This Shell pump, by then about thirty-five years old and still with a glass globe at its top, was to be found in St Cuthbert Street, Kirkcudbright, in the county of Dumfries & Galloway. The striped object appearing to stick out of it would be the barber's pole for the gents' hairdressing salon beyond it. (Chris Barker collection)

Petrol stations had been around for less than ten years when some places tested automatic vending pumps for the first time. Mortimer's Service Depot in Cleave, Somerset, experimented with a standalone pump painted white in 1927 – no doubt to show up better at night after the attendant had gone home. (Dafyn Jones)

An evocative image of night-time in the capital during 1934 is beautifully captured as a Talbot car pulls into a petrol station where the attendant is ready with hose in hand. The illuminated glass globes atop the pumps stand out well, as do the clock-style dials beneath. (John Kluge)

The irony of delivering cans of petrol to outlying areas meant it was a job usually best performed by a horse and cart when this photo was captured in 1910. The way men are gathered around the car, their bicycles parked against a bench and 2-gallon cans left lying around suggests that Pratt's was supplying petrol for an organised time-trial event. (Mike Jeavons)

Under construction somewhere in Canterbury, Kent, during the 1920s is how this photo is identified. Certainly, the Redline pumps, with their triangular glass tops, make an impressive island alongside the Texaco oil cabinets at this stage of the project. This would have seen the installation of underground storage tanks before the rest of the petrol station could be built. (Chris Barker collection)

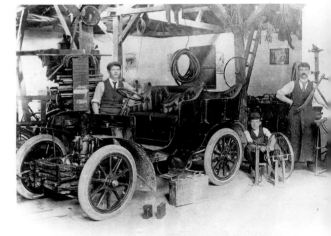

Hundreds of blacksmiths, cycle shops and hardware stores around the country adapted their businesses to meet the growing needs of the early motorist at the beginning of the twentieth century. Here, the men of Davis Garage in Ringwood, Hampshire, show off the variety of work they can carry out in 1904 including repairing cars, building bicycles and supplying petrol. (Ed Coldrick collection)

Steam-powered vehicles were the only ones considered strong enough to haul hundreds of 2-gallon petrol cans over any distance in the years before the First World War. Taken in 1911 near Anglo-American's Hammersmith depot in west London – Pratt's Motor Spirit was its key brand name – this wagon was built by Yorkshire Traction while the trailer appears to be a converted horse-drawn cart. (John Lea)

Petrol stations sprang up quickly around the country during the 1920s and were often adorned with advertising hoardings and signs. In the late 1920s the Council for the Protection of Rural England complained many were becoming unsightly and should blend in more with their surroundings. One solution was to make them more rustic, some having thatched roofs. The Haldon Thatch service station, with its row of five pumps, at Kennford near Exeter is a prime example. (Author's collection)

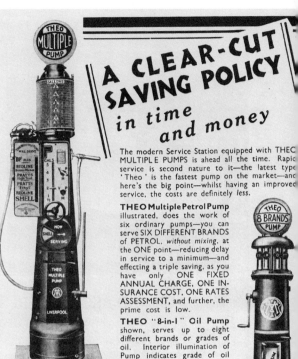

The Cornish village of Chacewater is just a few miles between Redruth and Truro. In 1934 one of its pubs, the King's Head, boasted a row of four pumps. The Essolene pump appears to be distancing itself from the others but in fact is placed to allow enough space for a car to enter and exit the garage behind it. While the pumps disappeared long ago, as did the 1/5d (7p) per gallon price disc, the pub still exists today. (Chris Barker collection)

This 1930s image of Wadham Brothers Garage in The Avenue, Southampton, shows an eclectic range of pumps. A dealer in Morris cars at the time – it would go on to sell Austin and British Leyland vehicles in the years that followed as part of the Wadham Stringer Group – BP, Redline and Shell brands were among the choices offered to customers. The octagonal hut in the centre of the island provides cosy cover for the attendant. (Stewart Would)

Dealer for Rover, Humber and Trojan cars, W. Foster Ltd of Amersham was a well-appointed place offering Texaco, BP, National and Shell products in 1930. The buildings appear to be quite new and may well have been built in conjunction with the Metroland housing schemes to create a new suburbia to the north-west of London. (Chris Barker collection)

2 WARTIME RATIONING —
JUICE IN A JERRYCAN

Unconvinced by Prime Minister Neville Chamberlain's 'Peace In Our Time' declaration in 1938 after his talks with Germany's Chancellor Adolf Hitler, Britain's oil industry prepared for war.

Believing they would be immediately nationalised at the start of any hostilities, the companies voluntarily offered to pool their resources instead of competing with one another – a move that paid off against being brought under government control.

The pooling arrangements included the companies' seagoing tankers, oil refineries, road and rail tankers as well as their employees.

It was only a matter of days after the government declared war on 3 September 1939 that petrol became rationed and the pool plan implemented. It was led by the government's Oil Control Board, which decided the price of petrol and its octane rating.

The first 'motor spirit' ration coupons could be collected from Post Offices or Inland Revenue offices for use from 16 September 1939 when measures came into force. Motorists had to take with them their registration documents and would be given two ration books, one marked first month, the other marked second month, and they had to be used within that time span or they would expire. This was designed to prevent hoarding of coupons – but not hoarding of petrol.

Each coupon in a book allowed you a gallon, although this might change as the war continued. The number of coupons received depended on the size of a car's engine, as shown on its registration document, and an owner's circumstances. People were restricted to about 100 miles' worth of petrol each month depending on their needs and their vehicle. Typically, a car owner could be limited to purchasing between 4 and 10 gallons a month at 1/6*d* (7.5p) per gallon.

There would be two kinds of pool, one for private use and the other for commercial, the latter being dyed red. The penalty for its misuse – putting it in private cars instead – was imprisonment. Fuel branding disappeared and at many sites, the glass globes atop petrol pumps bearing the oil companies' emblems were either removed, blanked out or overwritten with the word 'pool'.

Thanks to the pooling arrangement and the strict rationing, the price of legitimate petrol supplies held reasonably steady throughout the war at around 2 shillings (10p) per gallon.

Siphoning from a car's tank became an easy way of stealing petrol and enabled black-market supplies to thrive. It became common for tins of petrol and coupons to be traded for other short-supply commodities as well as cash.

This seems a drastic measure to ensure customers pay for their petrol! This scene of an Army field gun crew ready to fire on a visiting car was taken at Arter Brothers garage on the A2 at Barham, halfway between Dover and Canterbury, Kent, in 1941 during anti-invasion exercises. Only a few miles from here a railway tunnel housed Britain's largest gun regularly fired at the Germans in France during the Second World War. Arter Brothers garage still operates from the site and serves Gulf petrol. (Author's collection)

Petrol branding disappeared as soon as rationing was introduced, which meant that not only pumps were marked as 'pool' but road tankers as well. This vehicle belonged to BP but has been repainted to reflect the change when the oil companies shared resources and petrol during the Second World War, largely to avoid being nationalised. Note the edges of the mudguards, tail board and rear lights have been picked out with white paint to heighten visibility in the dark. (© BP plc)

Petrol was the first commodity to be rationed after war was declared on 3 September 1939. Within a couple of weeks, motorists had to collect ration books from Post Offices or tax offices to continue to obtain petrol. A differently coloured set of coupons was supplied depending on the engine size of your car or motorbike. These examples come from the Old Airport Garage Museum, Rochester, Kent. (Author's collection)

Much of Britain's precious reserves – nearly all of which were imported into the UK by ships from the Middle East – were needed for military and essential users, forcing many owners to lay up their cars 'for the duration'. In any case there was little incentive to use a car and pleasure trips were strictly forbidden.

By the summer of 1942 even the basic civilian petrol ration was ended as supplies dwindled and only official users could have any. Only the emergency services, bus companies and farmers were allowed any fuel. An exception was made for the travelling fairs taking their machinery around the country as part of the 'Holidays at Home' campaign. The idea of this was to raise morale and prevent people from travelling away for any kind of summer holiday, using up resources themselves. Not only were the fairs given unlimited petrol but they could summon local firemen to help set up and dismantle their equipment.

American soldiers filling up thousands of jerrycans somewhere in Normandy after the 1944 invasion. A line of petrol tankers disgorge their cargoes into a field full of cans for onward transport to the battlefront. The entire Allied surge to Germany depended heavily on a ready supply of petrol and when there was a shortage of cans to carry it, battle plans had to be redrawn. (www.thinkdefence.co.uk)

In this Ministry of Information photo, children play in the road while others help a GI stopping for petrol at the village pump in Burton Bradstock, Dorset, in the spring of 1944. The US Army had taken over use of the petrol station by this time. The soldier, Edward Shea of Dorchester, Massachusetts, shows a couple of the children around his vehicle while another wipes his windscreen and a fourth cranks the ageing pump. The cows, meanwhile, wend their way back to a nearby field. (Bridport Museum Trust)

Fuel supplied to approved users was dyed, and its use for non-essential purposes was an offence. One person who came unstuck was the actor, composer and musician Ivor Novello, who was jailed for four weeks after being caught misusing petrol coupons.

In general, cars were still a luxury only the middle and upper classes could afford. When the decision was made to ban pleasure motoring, there was much opposition from the well-heeled, as well as from shopkeepers, guest houses and hotels in seaside towns — at least the ones that weren't in military zones — because they relied on day trippers and tourists for a living.

Any parked car, laid up or otherwise, had to be immobilised to prevent it from being used by an enemy — usually taking the rotor arm out of the distributor was sufficient for short spells, but in the longer term it was wiser to take off the wheels and stand the car on blocks. Perhaps it was this measure that gave rise to numerous 'barn finds' when cars started coming out of hibernation years later!

Hanging out the flags when the end of petrol rationing was announced in late May 1950 are staff, customers and an Esso delivery driver at Higgs & Niblett's premises in Coleford, part of the Forest of Dean area in Gloucestershire. Until then, motorists had been limited to 100–200 miles' worth of petrol each month since September 1939. The end of rationing heralded a mad dash to the pumps for many – the owner of the Vauxhall car here not losing any time in filling up. (Graham James via www.sungreen.co.uk)

One option for a civilian driver to get around without the need for petrol coupons was to convert his or her car to run on coal gas. A large balloon, fitted on a car's roof, could be filled with enough gas for 20 miles. Conversion was expensive and not very efficient. Public transport experimented with the idea as well and many buses towed trailers that generated gas as they went along. There was considerable power loss with gas, so if a bus was going uphill, passengers might have to get off and walk to the top before re-joining it. Worse still, they could have been pushing the bus!

Meanwhile, petrol pump manufacturers turned over their factories to making munitions and equipment for the war effort. The notable exception was Beckmeter, which kept up production for the War Department, supplying them for refuelling lorries on RAF and Army bases around the country.

During the early part of the Second World War, the British Army relied almost entirely on thin metal cans of petrol and diesel with which to fuel its carriers, lorries and tanks. It wasn't until the war shifted to North Africa that its inherent weaknesses were revealed. The standard issue cube-shaped, tinplate 4-gallon petrol can did its job reasonably well in the European theatre of war, but once the Eighth Army moved to harsher desert terrain, the conditions soon took their toll.

Having realised the need for a sturdier design of petrol can to help win the war effort, the British government set its suppliers the job of making millions of them using the design adopted (or nicked, depending on your point of view) from the German pattern. Here, a young welder is at work fixing the top cap in place – a date stamp on the can's side suggests the photo was taken in 1942. (www.thinkdefence.co.uk)

Fuel leakage and wastage in North Africa was extensive, reckoned by one leading general to be as much as 30 per cent. The cans, made with crimped or soldered seams, split when the expanding vapours within built up pressure, made worse by bumpy transport over many hundreds of miles or being stored at open fuel dumps in desert heat.

Clearly, fuel leaking in vehicles already laden with explosives and likely to be fired upon was an added danger to soldiers' safety and these cans quickly became nicknamed 'flimsies'. Putting the fuel into a vehicle proved an unnecessarily prolonged business, too. A spanner was used to get the screw cap off and a funnel was needed to pour it safely into the fuel tank.

Once empty, these unpopular single-use cans could be put to better use by cutting a large hole out of one side to create a cooking stove. A small amount of fuel poured into some sand in one of these Benghazi Burners would be enough to boil water for a round of tea – aka the barely bloody drinkable!

Showing typical Teutonic efficiency, it was the Germans who had come up with the solution in 1937 while secretly rebuilding their armed forces. Properly known to the Germans as the Wehrmachtkanister, this can was designed by Vinzenz Grünvogel, chief engineer of the firm Müller of Schwelm. This larger 20-litre can was not only stronger, thanks to its indented sides, but its three hollow handles created an air pocket allowing the petrol some space for expansion. A flip-top lid and spout enabled the contents to be poured more carefully without spilling.

These welded jerrycans could be stacked on top of each other more easily without the lower ones being crushed under the weight. An impervious layer inside enabled them to be used for carrying either water or fuel.

Such was its military value to the Germans by early 1939 that Field Marshal Hermann Goering – Adolf Hitler's deputy and head of the Luftwaffe – personally intervened when three jerrycans went missing from Berlin's Templehof Airport. Two friends, a German engineer and his American colleague, had built a car in which to travel to India and needed the cans for the journey. They crossed the borders of ten countries before a special flight arrived to 'repatriate' the engineer. One of the cans stayed with the car and eventually found its way back to America. Displayed to the US Army's top brass, there would be little interest in further developing the jerrycan until it joined hostilities after Pearl Harbor in December 1941.

Opposite: Little encouragement was needed to bring drivers to petrol stations once the end of rationing was announced in May 1950 – but the staff at Lex Garages in London put up a chalk board anyway. The 'private' label on the pump may indicate petrol for cars, as opposed to 'commercial' diesel for lorries and trucks. (Alan Graham)

PETROL
BOTH SIDES
NOTICE TO CUSTOMERS
SMOKING
PETROL OFF
RATION
COME AND
GET IT
LEX
PRIVATE
LEX
PRIVATE
LMB 216

SALES
PRINGFIELD GARAGES LTD.
PREMIUM
ICA
SHELL AND BP SERVICE
NNK 436
VACANCIES
NO PETROL

Meanwhile, Britain's Desert Rats had captured cans from the Jerries (Germans) and nicknamed them jerrycans, and they began using them instead of their own wherever they could. In 1941 a large quantity of jerrycans fell into British hands after the second battle of Benghazi and many of these were used to equip the vehicles of the Long Range Desert Patrol Group.

Back at home, the government had several suppliers, including Vauxhall Motors, making the British jerrycan in their tens of thousands. The American government, now realising the design's potential, adapted the idea slightly and made their own version.

With the war in full flow across France, Belgium and Holland after the Normandy invasion of June 1944, it was clear that jerrycans could be used for many things other than holding petrol. Many were used as makeshift stepping stones across muddy ground, for example.

Opposite: The owner of this Sunbeam Talbot would have had a job finding petrol in late November 1956 as supplies quickly ran out following the announcement of rationing during the Suez crisis. Springfield Garages at Palmers Green in north London was unable to help on this occasion. (Chris Barker Collection)

Examples of American and British jerrycans at the Old Airport Garage Museum in Rochester, Kent. Although both cans were made years after the Second World War, the designs were unchanged. The British version, on the right, stayed closer to the original German pattern with its more robust flip-top cap and indentations on the side panels. The US model opted for a cheaper cap. (Author's collection)

At one point, the Allied armies reckoned to be short of 3.5 million cans and plans to push forward towards Germany had to go on hold while replacements were found. However, one million cans were returned thanks to French schoolchildren, who had been incentivised with the promise of pocket money for each empty one they found.

Although the Second World War ended in 1945, it would be another five years before petrol rationing finally came to an end. The Labour government of the post-war austerity era maintained that rationing was still necessary because the country didn't have sufficient US dollars, the global currency used to pay for petrol supplies.

Good news arrived on 26 May 1950 when the Minister of Fuel and Power, Philip Noel-Baker, announced in the House of Commons that petrol rationing was being abolished. Two American companies had agreed a deal to supply more oil in return for buying British goods.

Under a deal brokered earlier that month, the Standard Oil Company (Esso) and the California Texas Oil Company would be paid in British pounds for supplies. In turn, they would invest that money in British equipment, services and oil tankers.

The announcement was followed by a mad rush to petrol stations as thousands of motorists ceremonially tore up their ration books.

The government estimated a rise in petrol consumption of one million tons a year. About 430,000 tons of this was to be supplied by the American companies. The rest would come from newly expanded refineries in Britain but quality issues would remain for another couple of years or so while additional refineries were built at Southampton and in Cheshire.

The Treasury was expected to collect an additional £26m a year in fuel duty and from savings in administration costs. More than 2,000 civil servants who ran the fuel rationing system found themselves out of work or being diverted to other duties.

Jerrycans, now often made of plastic, continue to be used around the world but in far smaller quantities and generally only by forward-facing troops travelling over shorter distances. There are plenty more tucked away in sheds or on the backs of off-road vehicles, too!

3 | FUELLING 1950S GROWTH – AND A SUDDEN STALL

Having agreed to help the British Government bring an end to rationing in spring 1950, Esso lost little time in trying to capture a bigger share of the country's petrol station network.

Before long, the company became the first to introduce a scheme where the nominally independent retailer signed up to selling only Esso products for several years, in exchange for a better price on wholesale supplies.

In little more than another twelve months all of the major oil companies in Britain had made similar 'solus' arrangements, as they became known, with the majority of their customers. Gradually, rows of pumps of many shapes and colours would be replaced by those of similar size and one uniform livery.

Not everyone was happy – some retailers who wanted to stay independent banded together to complain to government about oil companies refusing to supply unless they signed up to a solus deal. Castrol, the independent supplier of lubricating oils, found itself losing ground to corporate-labelled alternatives and ran a campaign to encourage people to complain if their products were not being offered on solus forecourts.

It would be the mid-1960s by the time this issue was finally addressed by the Monopolies Commission and even then, because the oil companies pleaded poverty and predicted financial ruin, the body effectively waved away the independents' plaintive cries of oppression.

During the mid-1950s there were three established big brands and a couple of relative newcomers controlling much of Britain's petrol supplies: Esso, which had also owned Cleveland since the mid-1930s; Shell-Mex and BP, who had shared distribution networks for some years already, also operated National Benzole and Power brands; Regent, which was owned by Texaco and Chevron; Vacuum Oil, which changed its name in 1955 to Mobil Oil; and Fina Petroleum, which had arrived on UK shores in 1951 after buying distributor Cities Service Oil. The company became Petrofina in 1957.

Growth in car ownership in Britain had jumped by 250 per cent in just a few short years as modern-designed vehicles finally started rolling off the nation's assembly lines. Affording a car became easier too once the government allowed hire purchase schemes to be offered to eager buyers for the first time – but high rates of purchase tax, very often one-third or more of the price of a new car, kept the market in check.

More cars on the roads meant more business for the petrol stations, of course – as well as more work for mechanics to service and maintain them. Reliability was not necessarily built into a car in the way it is today!

The cost of petrol had risen exponentially since 1945, still being controlled by the government until rationing ended, and by 1953 the cost of a standard-grade gallon of petrol stood at 4/2d (21p), of which 2/6d (12.5p) was taken in duty by the government and 4d (2p) as profit for the petrol station.

Motoring's golden decade was brought up short in late November 1956 when the Suez Crisis – triggered by Egypt's President Gamul Abdul Nasser taking over control of the Suez Canal four months previously – saw petrol rationing reintroduced.

Oil tankers from the Middle East relied on passing along the canal for a shorter journey to their European destinations rather than take a longer, and more perilous, voyage around the coast of Africa. Consequently, British oil refineries were in danger of running dry.

Initially expected to last for four months, rationing actually lasted nearer five months. Petrol coupons, which had been carefully stored since the end of the war, reappeared and most people were restricted to just 200 miles' worth of petrol each month. Given that the majority of cars could deliver no more than 30mpg, this worked out at about 7 gallons.

PROPOSED SITE.

Set among terraced houses of Waterloo Road, King's Heath, Birmingham, was the Waterloo Garage, pictured in 1953. The print has been used to identify the positioning of a proposed sign, perhaps heralding the arrival of new pumps as well? (Andy Maxam, maxamcards.co.uk)

News that petrol would be rationed in late 1956 as a result of the Suez Crisis saw queues quickly build up at petrol stations around the country. Pumps soon ran dry. In pole position among the vehicles on this forecourt are a split-windscreen Morris Minor convertible, a Vauxhall Wyvern, and an Austin Counties saloon. Further back, a lorry driver and his mate try to beat the boredom of waiting. A little way behind them is a chauffeur-driven Rolls-Royce. (Brian Guy)

Business users were limited to an additional 100 miles' worth a month, while farmers, the clergy and essential local authority staff were allowed a more generous 600 miles. Meanwhile, doctors, midwives, vets and disabled people could have whatever petrol they needed.

It would take nearly three weeks after the announcement for rationing to be fully implemented – a gap that was widely criticised for giving people the chance to hoard illegal supplies. In the meantime, petrol stations

The green and yellow livery of BP would have shown up well in this 1959 photo of its new service station at Bagshot, Surrey. There has been considerable effort to establish BP's corporate look around the site as it shifted from being sold alongside Shell and National. The signs to the right indicate twenty-four-hour opening, two-stroke being on sale and agency cards, aimed at company fleets, were accepted. Beyond an array of cars of the era is a Tate & Lyle Bedford lorry. (© BP plc)

were asked by the government to voluntarily restrict daily sales to a couple of gallons per customer – which many did.

Perhaps predictably, panic buying ensued and queues quickly built up at forecourts around the country.

One BBC reporter was told by a petrol station owner in Denham, Buckinghamshire: 'We are almost afraid to serve our regular customers. When motorists saw a car being filled they stopped and waited. In five minutes we had a queue of fifty cars waiting – and we had to turn them all away.'

By the time rationing started in mid-December, roads were deserted and petrol stations had closed across much of the country. Petrol was said to be virtually unobtainable in the centre of London but had the beneficial effect of reducing traffic in the capital by two thirds.

Two classic 1950s Fords, a Consul, left, and a Prefect, are fuelled up by female attendants at BP's Roundabout garage at Greenford in west London during the late 1950s. BP Super sells for 4/7d (23p) and BP Super Plus for 4/9d (24p) per gallon from pumps beneath lighting panels, which would afford minimal cover in the rain. Inside the circular shop two more staff wait to sell a range of other things that, judging by the sign at the top of the window, include Player's cigarettes. (© BP plc)

Set beside a roundabout to serve the busy A20 and A205 South Circular arterial road around the perimeter of London is the Clifton service station in Sidcup Road, Sidcup. The site continues to trade today under the Esso banner – with the main building still in situ – but back in the 1950s, this usually busy place boasted a line-up of no fewer than ten Beckmeter pumps, beneath a translucent canopy. (Chris Barker collection)

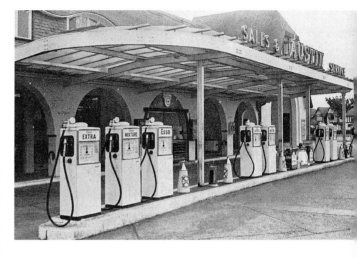

People were outraged as pump prices rose sharply at this time to around 6/- (30p) per gallon, the main suppliers saying they had to make up for lost revenue somehow. The government went as far as adding a surcharge to prices to help compensate the trade for its loss and the cost of bringing oil around the southern tip of Africa. However, by the time rationing ended, the oil companies still reckoned they were £4 million down.

The Suez Crisis had a major effect on British industry as a whole – the government had imposed a 10 per cent reduction in fuel supplies to industry, forcing many factories on to short-time working. Commercial travellers, used to carrying their wares by car, had to find other ways of reaching their customers.

This spell of petrol rationing came to an end on 26 May 1957, just before the Whitsun weekend, when Paymaster General Reginald Maudling announced to the House of Commons that restrictions would be lifted right away because petrol stocks were 'at a satisfactory level'.

Opposite: A Derbyshire-registered split-windscreen Morris Minor Traveller is filled up a by a young attendant, resplendent in tie and neat overalls at a BP petrol station on the A3 Kingston bypass in Surrey, sometime in the mid- or late 1950s. Built in the 1920s to connect Portsmouth with London, it skirts around the town of Kingston-upon-Thames and has become one of, if not the, busiest non-motorway roads in Europe. (© BP plc)

The Guardian newspaper described events the following day:

Motorists drove to the petrol pumps and filled up with coupons that would be useless or found they were no longer needed.

'All off ration,' chalked one petrol station attendant. 'Come and get it.' The motorists queued to do so. Some stations ran out but the shortage will be temporary. There are enough stocks, it is confidently said, to meet demand, though motorists may now have to use a little patience to get their tanks filled.

Tonight almost everybody is delighted. The government is proud of itself. The opposition feels it can take all the credit for its steady pressure. The oil companies are purring. The motoring organisations are happy. The garages are looking forward to increased sales. The secondhand car dealers are sure prices will go up.

Hotel keepers feel they can now be reached by customers with money to spend on their new, beautiful and lengthy menus. Investors in garages, oil and automobile shares feel richer, and the ordinary motorist is looking at long-distance maps.

Set on the Great West Road at the junction with Ealing Road in Brentford was this imposing place to fill up. In 1955 the Dome Garage, operated by Lex Garages Ltd, offered the full variety of Regent products. The Dome was more than just a filling station, being part of a large complex offering a trendy snack bar and a hugely popular ballroom. (Chris Barker collection)

Administering this round of petrol rationing had cost the government about £20,000 a week and relied heavily on driving test examiners to carry out that work. During that period, tests had been summarily cancelled and learners were allowed to drive unaccompanied.

WS 23 THE CLOCK TOWER, WARSASH A TUCK CARD

A Tuck postcard published during the 1950s of the local clock tower has recorded not one but two neighbouring petrol stations in the Hampshire riverside village of Warsash – but neither place appears to have any customers. Cleveland Discol and Benzole are on offer at around 4/9d (28p) per gallon in the foreground while the BP Shell Mex range can be found further back across the road. (Chris Barker collection)

Left: Motor oil was sold in bottles for many years until the late 1950s and early 1960s. (Author's collection)

Below: Until the late 1960s, most petrol pumps had clock style faces showing how many gallons of petrol were being pumped into your car. These examples are on display in the Old Airport Garage Museum, Rochester in Kent. (Author's collection)

This picture of the petrol station in the Cornish village of Grampound, 8 miles from Truro, was taken around 1950. The Shell Mex tanker driver, wearing his peaked cap, has paused on his delivery round to pose with the proprietor. At the end of the 1960s, the petrol station was demolished as part of a nearby river bridge widening scheme and rebuilt further back. It became a car sales centre from 2011. (Grampound With Creed Heritage Project)

Thought to date from the late 1950s, Lambert Motors of New Road, Peterborough, was an early acquisition for Frank Smith, who went on to own a number of garages in the region. At this time Lambert Motors was a Rover/Land Rover dealership but took on the Vauxhall franchise for Peterborough in 1969. Lambert's remained part of Smith Motor Group until the mid-1970s, when it was demolished to make way for a new road. (Smith Motor Group)

Captured in all its art deco splendour, even twenty or so years after construction, Lambs Ltd was a landmark petrol stop, service garage and car sales centre along the Southend Road in Woodford, Essex. In this early 1950s view, not only do we have a battery of multi-branded petrol pumps but also a separate island for Castrol oils, for blending either into two-stroke or straight lubrication. Southend Road was one of four Lambs outlets, later becoming a major dealership for Standard and Triumph cars. (Chris Barker collection)

The grass bordering the front of the forecourt has just been cut judging by the lawnmower abandoned at the kerbside of Littleton Garage on the A361 at Semington near Trowbridge in Wiltshire. No doubt a customer had called in for petrol dispensed from one of the stout-looking Cleveland pumps. Although petrol sales ceased here long ago, the garage today is an independent specialist selling and servicing Land Rovers and Range Rovers. (Chris Barker collection)

When this photo was taken of Martin Walter's premises in the mid-1950s, Cliftonville was the smarter part of Kent seaside resort Margate. The Esso pumps, flower beds and spacious semicircular showroom emphasised this grandeur. Martin Walter was the parent company of Dormobile, the famous camper van builder in Folkestone. The Cliftonville site was sold to Ford dealer Invicta Motors around 1971 and redeveloped in the 1990s. (© Thanet District Council SEAS Heritage Collection)

Occupying sizeable premises along Pershore Road, in Selly Oak, Birmingham, during the 1950s, when this image was taken, was Taylor & Grainger's Austin dealership and Cleveland petrol station. Swing arms would carry fuel pipes above the pavement to waiting vehicles. (Andy Maxam, maxamcards.co.uk)

4 FUEL FOR THE MOTORWAY AGE

Exactly forty years after the debut of Britain's first dedicated petrol station came the opening of the M1, the country's first motorway and, with it, the first service area.

Much has been written elsewhere about the quality – or lack of – of motorway services' cuisine, but their adjoining petrol stations have barely merited a second glance, other than to vent about extortionate pump prices.

The M1 saw the first traffic pounding up and down its new tarmac between London and the Midlands in November 1959 with drivers, and their vehicles, able to pull over for a breather at the Watford Gap services.

During the build stages – the road took only eighteen months to create – planning the service areas had become an afterthought and it wasn't until September 1959 that the Ministry of Transport ordered its appointed contractors to drop everything and get the job done.

Blue Boar, which started out as a vehicle recovery business and ran a nearby petrol station on the A5, had won the contract for the Watford Gap services. It only just managed the 2 November opening date after being beset by problems with late deliveries, bad weather and time-consuming earthworks.

Further up the motorway at Newport Pagnell, rival operator Blue Star, working in tandem with restaurant chain Forte, had to face up to selling petrol and diesel from jerrycans until it could put something more permanent in place.

Although officially opened in June 1987 by Prime Minister Margaret Thatcher as part of the last stretch of the M25 motorway around London, BP's site at the South Mimms services had already been operational for seven months serving local traffic. At the time, the petrol station was the largest of its kind in Europe, equipped with no fewer than twenty-six pumps offering mainly two- and four-star petrols with only a couple set aside for unleaded and diesel. (© BP plc)

It wouldn't be until autumn 1960, about a year after the M1's opening, that the two service areas would be fully up and running with finished buildings and gleaming petrol pumps. By then, it had been realised that service areas as a whole were seriously lacking in being able to meet their demand. The Ministry of Transport had predicted that one in seven vehicles would use them but, in fact, one in four were choosing to turn off for the three Rs of rest, refreshment and refuelling.

In those days, there were only 4.5 million cars on Britain's roads, a fraction of today's number, and the motorway itself was unlit with only a grass verge dividing the up and down carriageways. Clearly, not a place where you would wish to linger if you didn't have to. Plenty did though in the early years as the wheezy cars of the era, more used to pootling around towns at a sedate pace, weren't designed for extended periods of high-speed motoring. Either their radiators or big ends were usually the first components to give way under the strain – keeping the motoring organisations and recovery trucks in honest work.

Speed limits weren't imposed on the motorways until 1965, by which time much had to be done about deaths on the nation's roads. The new 70mph maximum limit reduced fatalities on motorways and A roads by 20 per cent in a year.

Despite solus arrangements in place at petrol stations across the country since the mid-1950s, motorway service areas offered several brands of petrol when they first opened. Government lobbying in the interest of fairness by the AA and RAC insisted upon more choice for motorists, thus giving us a now rare sight of three names beside each other on a snowy day at Blue Boar's Watford Gap services in 1960. (Motorway Services Online)

In the interests of fairness, providing customers with some sort of choice and after intense lobbying from the AA and RAC, the first motorway petrol stations were made up of serried ranks of different branded pumps. You could pull up alongside islands of Shell pumps, a line of Fina ones or even a quartet of National dispensers, for example. All of this represented a big departure from the solus operating principles that had become near universal everywhere else since the mid-1950s.

It wouldn't always be like this though. As motorways, and the service stations, became more numerous with sites frequently changing hands and brands, so the larger companies took over an entire area's fuel operations, while others pulled out altogether.

A case in point, perhaps, arose when BP took on the development of the entire South Mimms services on the northern stretch of the orbital London M25 in the 1980s. It was the first in the country to be built with private money – until then the Departments of Environment and Transport had been responsible for the funding and development of motorway service areas.

BP leased out the main facilities to Trusthouse Forte, who branded South Mimms as a Welcome Break area and chose to run the truck stop and petrol station – when it officially opened in the summer of 1987 this was the largest of its kind in Europe. Later on, the Department of Transport bought the site from BP and leased it back under a 125-year agreement.

Motorway service areas were seen as so innovative when they first opened, they quickly became social gathering points in their own right. This postcard of Keele services on the M6 takes advantage of that – and highlights the petrol station area in doing so. (Author's collection)

Previously, South Mimms had had a colourful history. At the junction of the A1 and A6 roads serving London, it was known as Bignells Corner after a garden centre there. Over time, a large truck stop sprang up at the Beacon Cafe alongside an Esso-run motel. The area became infamous for the exchange of black-market goods and prostitution, but both had been quelled by the 1960s.

Throughout the sixty years of development of Britain's motorways there have been continual complaints over petrol pricing. Certainly, from the mid-1960s onwards, motorists have been saying they have paid through the nozzle for the dubious pleasure of filling up at a motorway service area.

Even today, and despite promises of a crackdown by successive governments, pump prices have been historically much higher than on other carriageways, and certainly when compared with supermarket forecourt prices.

In December 2018, *AutoExpress* reported:

Recent figures from the RAC suggest motorists topping up at a motorway petrol station pay up to 15 pence per litre more than elsewhere. Motorway stations argue the reason their prices are higher is that many are open 24 hours a day and offer more services than a regular forecourt. Motorway fuel stations also pay high rent prices for the buildings they operate.

In more remote areas, fuel is often more expensive because of the higher transport and supply costs, but according to RAC fuel spokesman Simon Williams, this doesn't apply to motorway stations: 'We can see no reason why motorway fuel should be so much more expensive. In fact, arguably it is much easier from a delivery point of view than it is getting fuel to urban filling stations.'

5 IT SHOULDN'T HAPPEN TO A PUMP ATTENDANT!

It's easy to paint a blissful picture of life as a petrol pump attendant in the pre- and post-war eras of the golden days of motoring.

We have a subconscious image of a reasonably friendly middle-aged man or woman clad in smeary overalls, or knee-length white coat, diligently getting the pump running, perhaps swinging its overhead arm so the nozzle can reach your car's tank before filling up with a leading brand of fully leaded 'super' grade petrol.

Customers would receive, and appreciate, good service, enjoy some cheery banter with this person – who might well have been a member of the family who owned the place – and all without any fuel being squirted down the side of the car.

The truth could be different, of course. *Autocar* magazine interviewed one petrol station proprietor in the more trusting times of May 1965 to learn something of the reality of life manning the pumps.

Mike Bailey, manager of a Regent outlet in the Berkshire village of Pangbourne, suggested that many drivers got the service they deserved. Realising the business was becoming ever more competitive since petrol prices were no longer fixed, he wanted to ensure his premises were as efficient and attractive as possible – despite his customers' best efforts.

Autocar reported:

Garages have relied on trading stamps, giveaway offers, vending machines and clean toilets to attract new customers. What is the result?

Take washing facilities first. Mike has had to change from linen to paper towels because the linen ones were being continually cut in half; soap is stolen almost as fast as it can be put by the basins – bottled soap is simply poured away; piping is torn away bodily from the walls and the paintwork has to be renewed once a fortnight.

Many of a service station's facilities are free to customers but if they are abused they can cost the long-suffering proprietor a considerable amount of money. Thus after the fourth tyre gauge had been broken – they are £8 a time – by being thrown down or driven over, it was replaced with a dial type gauge.

A brand-new coffee vending machine had to be repaired within days of arrival after a customer had driven into it, shifting it 8ft across the forecourt. A vacuum cleaner supplied for customers' use didn't come through unscathed either. The magazine continued:

The usual game is to use the vacuum cleaner and then reverse over the rubber pipe. It is a pretty strong one, but eventually something has to give. Then, of course there are people who drop the washer brush into the grit so that the next person to come along covers their car with clean bright scratches.

New tyres were routinely chained to outdoor display stands. 'Absolutely necessary,' said Mike in the article. 'They would be gone within hours if they weren't. We've had three break-ins recently. One mob made off with all the oil and the other two did the cash box. Of course, that's not counting the times the slot machines have been jemmied at night.'

Most customers were thoughtful and considerate but others would empty their ashtrays while the tank was being filled or scatter their sweet wrappers to the four winds. The two most awkward types of customer were those who claimed to have no money on them to pay for their petrol – after their tank had been filled, naturally – and those who demanded impossible repair jobs.

Too many cars were still on Britain's roads with badly worn and positively dangerous tyres (laws around minimum tread depth had yet to come into force at the time). Point this out to a customer, ran the article, and the reaction would often be that you were trying to sell him something that wasn't needed.

Bromley Road, Catford, in south London is where this jack-the-lad attendant worked in 1967. A non-corporate uniform style of flat cap, old jacket and white overalls suggests a more traditional approach to forecourt fashion. (Ed Coldrick collection)

Autocar finished off by saying: 'Petrol stations do try to improve their services but this cannot be economic if they have to pay out regularly to make good the damage done by vicious or careless patrons. Failure to respect garage premises and their attendant services harms everyone in the long run.'

The lot of an attendant had hardly improved five years later when one felt forced to put pen to paper to sound off to the AA's members' magazine *Drive* in New Year 1970. T.J. Winspear of Irlam in Lancashire wrote:

I have been a pump attendant for six years. If I tell you how motorists, both male and female, make my job more difficult, it may answer some of your readers' questions.

This is what they do, they pull up on the wrong side of the pumps beyond the reach of the hose, smoke while I'm serving, ask for four gallons and, when the tank overflows at two, blame me for spilling it!

They suggest that a drop of petrol on the wing should be wiped off when the car is so dirty my overalls get filthy while leaning over to clean it.

They grumble at me for any price increase, as if it went into my pocket. They expect me to know what sort of oil they use, even if they have never called before. They keep the windows up and talk through the glass – I am not a lip reader!

When they come to see the boss about a repair, they leave their car blocking the pumps. They hand me a bunch of keys and expect me to know which one opens the petrol cap. They ask what sort of petrol I would recommend – and then ask if I'm sure.

They let children and dogs out of the car without keeping them under control – one child was nearly run down on the forecourt.

These are just a few examples, but I still enjoy my job.

Filling stations are rarely noted for being glamorous places. Cars and their passengers spend no more time there than they need to while on the way to somewhere else. Besides, exhaust fumes, oil and the inherent risks associated with large underground tanks full of flammable petrol and diesel are enough reason not to linger.

In more modern times, customers might stick around long enough to buy some wrapped flowers, a paper, milk, a sandwich swathed in plastic or a bar of chocolate, but that's about it.

As far as fashion goes, the 1970s was the decade that good taste forgot and this young lady's attire only emphasises that. Flared trousers with crimson zip-up dust coat doesn't make the best of pairings! She would have been easily spotted on the forecourt in the days before hi-vis jackets. Taken in 1978 at Furnham Road petrol station in Chard, Somerset, the Elf-labelled Avery Hardoll two-star pump may previously have been VIP branded. Judging by the Mobil canopy beyond, local competition was keen. (Mike Grant)

Service with a smile seems assured at Bishops Meadow petrol station in Wales' Brecon area, sometime in the late 1960s. The Shell outfits of white rugby shirt and yellow trousers seem far more practical – and a lot warmer -- than the hot pants and miniskirts provided by other oil companies to their forecourt staff. (Ed Coldrick collection)

Opposite: A 1965-registered Hillman Minx gets 4 gallons of Cleveland Discol from a young lady who, judging by the straight creases, may have only unfolded her new uniform from the packet that day! The site is thought to have been located at Margate Road, Ramsgate, where a second-hand car business now exists. (© Thanet District Council, SEAS Heritage Collection)

Bringing more customers into a petrol station and keeping them there long enough to extract some extra cash has tested the oil companies' marketing teams for many decades, especially since solus arrangements were introduced and the wholesale buying up of independent outlets.

We look at the special offers and giveaways elsewhere in this book, but another means of making petrol stations appear more enticing during the 1960s and '70s was to import the glamour.

Petrol stations converting to self-service, changing ownership or boosting a giveaway offer all needed promoting, so what better than to bring in photogenic young women to wow the mainly male motorists?

Clad in tight-fitting T-shirts, miniskirts and, latterly, hot pants, they would be on hand to minister to motorists' petrol-buying requirements, no doubt raising enough double entendres for a full-length *Carry On* film!

Cleveland updated its corporate look at the end of the 1960s and expected staff to wear this range of uniforms. The hats are probably the funniest things – the men get American copper's caps, straight from NYPD, while the women had to endure wearing surplus Russian airline stewardesses' hats that made them appear 4in taller. More in keeping were tunics, anoraks and sweaters. Ladies' minidresses and hot pants were for warmer days or when the petrol station around the corner hired in bikini-clad promo girls. (John Asplin/Author's collection)

Opposite: In less-enlightened times, comments about the lung capacity required by these young women to blow up their balloons would doubtless have come to the fore! Not so today, but we do know they were pictured after taking part in a balloon-blowing contest, representing EP and European petrol brands at the Essex County Show in Chelmsford in 1968. Pictured from left are: Terry O'Donoghue, 24, Patricia Walton, 20, Jane Cameron, 18, and Anne Dunhill, 19. (Chris Barker collection)

Nor was it unknown for rival outlets to recruit young women willing to disport themselves in bikinis as they served petrol in the warmer months – although one petrol station owner near Leeds discovered he could attract just as much extra custom by standing two or three tailor's mannequins beside his pumps.

Today, it would be a foolish forecourt owner who resorted to such sexist practices, but at the time the major oil companies thought little of it. One smaller company, European Petrol, hired a team of society girls to brighten up life at its new petrol stations around London during the mid-1960s. They were only expected to do an hour's work on opening day before handing over to more seasoned employees.

A trendy outfit like this wouldn't have suited every female pump attendant back in 1970 but it worked well for this young woman, although the boots may have been her own accessory. Somehow, the Stag Service Station seems an appropriate name for a place to be encountering such a stylish staff member. With five-star petrol on offer, it promised to be a high-octane experience! (© National Motor Museum collection)

London's swish Dorchester Hotel 'on' Park Lane saw the launch of European's Gold Seal petrol in the mid-1960s with the assistance of a troupe of society girls. The young ladies had to (wo)man the pumps for the first hour at every European petrol station selling Gold Seal, then priced at 5/6d (28p) per gallon. Turning an ankle, from left, were Nikki Phillips, 18, Roxana Lampson, 20, Carolyne Houston, 17, Olivia Smithers, 18, Moranna Cazenove, 20, Rosemary Walduck, 19, Fiona Hughes, 21, and Anne Dunhill, 20. (Chris Barker collection)

When Mobil launched its design award-winning cylindrical pumps in 1973, two hot pants-clad models were paid to drape themselves around one for media photos.

We shouldn't forget that promotions girls were widely used to promote a full range of consumer products in those days – not least of all cars, as demonstrated for many years at the annual Earls Court Motor Show in London. A few were required to bare nearly all to win the media's attention for some ranges of cars.

Much of this disappeared once self-service stations caught on and the forecourt staff became confined to manning an electronic panel controlling the pumps and taking payments inside a comparatively warm shop. A microphone on the desk would become the sole means of nagging at customers struggling with a temperamental pump or instructing them not to spill petrol about the place.

Once indoors, forecourt staff were soon expected to have a role in developing the shop from being an afterthought selling sweets, fan belts and oil to becoming the fully fledged 'retail theatre' experience that motorists are confronted with today.

Attended service isn't quite a thing of the past. It is still available on the Channel Island of Jersey, where family-run Falles operates at Airport Road, St Brelade and at its HQ at Longueville, St Saviour. In the 1970s and '80s Falles ran a fleet of up to 2,000 hire cars – easily spotted by the 'H' plate beside the registration number. In this 1979 scene, blue-uniformed representative Carol has a Ford Fiesta replenished by attendant Kenny in readiness for a customer. (Falles Ltd)

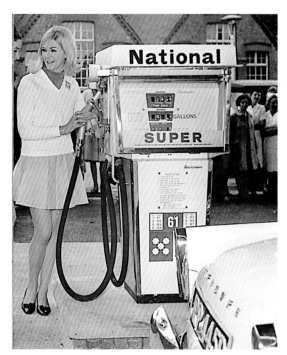

National's corporate uniform style had a more flattering look when this image was captured at the reopening of premises in High Street, Harborne in Birmingham, c. 1970. The new Beckmeter pump, offering five-star fuel at 6/1d (31p) per gallon, emphasises how petite the model is. (Tim Pearson)

Jim Clark, Formula One motor-racing world champion in 1963 and 1965, purchased the Northern Garage, by the A1, at Tweedmouth, in November 1966. The *Berwick Advertiser* pictured him soon after filling up a customer's Rover 2000. Jim told the newspaper he hadn't thought about using the place for racing-car development, nor did he have plans to retire from the sport as he had a full programme. Jim continued in motor racing until his death at the West German Hockenheim circuit in 1968, aged 32. (Author's collection/Berwick Advertiser)

Hot pants and waistcoat had become order of the day for some in 1972 as worn by a pump attendant at Moor Lane petrol station, an Esso retailer in Preston, Lancashire. (Preston Past & Present)

When the pumps have run dry on your forecourt there is little you can do other than read a book or start knitting, as these Hendon garage assistants happily showed for a press photographer during a tanker drivers' strike in 1956. The Army had to be called in to keep supplies flowing during the dispute. (Chris Barker collection)

Drivers of posh motors could be sure of special attention when they arrived at the new-look Swan Motor Centre, Oxford, in August 1973. Female staff were happy to disport themselves for the owner of this 1933 Rolls-Royce anyway – hopefully he put his quadruple Co-op stamps to good use too. (Joel Mutton)

At some time in their working lives, petrol pumps would have been tested by local councils' weights & measures departments to ensure they were delivering the correct amount of fuel. The location for this earnest pair of inspectors going about their work is unknown but it was captured in 1963. (Chris Barker collection)

Persuading Customers to Pull In

Left to right: With petrol stations recovering from the Second World War and growing numbers of cars taking to Britain's roads throughout the 1950s, competition was fierce among pump manufacturers. Avery Hardoll was keen to stress reliability and speed of delivery among the essential features of its dispensing equipment. (Chris Barker collection); Beckmeter pumps were made in south-west London and were a popular choice for Shell stations, among others, throughout the 1950s and '60s. (Chris Barker collection); A very curvy two-seater sports car pulls in for refuelling at a Cleveland petrol station in this October 1951 advert, which appeared in *Motor* magazine. (Author's collection)

Hear that tiger talk!
GRRRRRRREAT NEWS

NOW! 3 GREAT NEW PETROLS FROM ESSO

Here's the greatest motoring news in years. Esso have now made an entirely new range of petrols. Three great new grades to get the best out of every kind of engine. Now you know you can drive in to every Esso station in the country and get the perfect petrol for your car. It's as easy as that! ESSO EXTRA · ESSO PLUS · ESSO

(Esso)

Your engine's quick response to the accelerator *proves* you can't beat ESSO EXTRA for consistently high performance. And little wonder! For this magnificent petrol's unbeatable combination of 'extras' gives you quick starting, swifter warm-up, faster acceleration and extra power *when you want it* ! What's more, ESSO EXTRA's unequalled high quality and superb engine cleanliness give you more miles per gallon, outstanding anti-knock value and extra valve protection at all times. Try ESSO EXTRA today and prove it yourself.

ESSO EXTRA

THE FINEST PETROL IN THE WORLD

The introduction of a common standard for grading petrol, known as BS4040, was introduced in early 1967 to give us the two-, three-, four- and even five-star ratings. They would be familiar to millions of drivers for the next twenty-eight years before Britain moved to unleaded petrol in the mid-1990s. This was how the Esso tiger broke the news of the stars' arrival. (Author's collection)

Putting a tiger in your tank would prove to be a winning slogan for Esso for around twenty years after this advert first appeared in 1955. Although that phrase isn't used here, the connection with tigers had been made. (Author's collection)

NEW-CARS
AND NO-LONGER-NEW-CARS
GO NEWER,
LONGER
WITH SUPER NATIONAL

New cars need the very best petrol to perform properly, not-so-new-cars need it to bring back their youth. That's where Super National comes in! Everything is go-ahead about Super National. The people who buy it. The people who serve it. And especially the cars that use it. Super National promises any car fast warm-up, zip-away starting, and really eager acceleration. Try a tankful and see!

The attendant filling up this Ford Zephyr with National petrol in this 1962 advert must be thinking he meets some very strange people on his forecourt who want only to dance around and throw their arms in the air. Perhaps they are trying to tell him they wanted 4 gallons, not 6? Anyway, the then still relatively new blue and yellow National livery dominates here and the message that its petrol is good for both new and old cars is easily remembered. (Author's collection)

MOTOR INDUSTRY *for September* 1958   81

Throughout the 1950s, oil companies were keen to be sole suppliers to petrol stations across the country – bringing to an end a pre-war tradition of forecourts offering a number of brands and grades of fuel. National Benzole – as it was still known in this September 1958 trade advertisement – was working to attract more forecourts its way. Unfortunately, the Bedford Dormobile and a line-up of ageing cars beyond the pumps added little glamour! (Chris Barker collection)

Perhaps unusually in the more machismo world of 1954 and an apparent horror of women drivers, this advert for Regent TT petrol and Havoline oil places a woman in the driving seat of a sports car, drawn to resemble a Jaguar XK120. Her suited suitor looks on benignly in front of a pump that towers over him! (Author's collection)

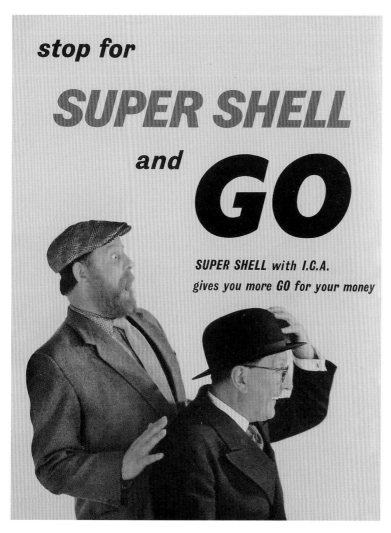

A simple yet timeless advert for Shell petrol dating from 1957. The image of the city and country gents looking on in surprise at a passing car could have appeared in much more recent years without being out of place. (Author's collection)

French-owned Total had been in Britain for only seven years when this colourful ad featuring a lantern-jawed lad driving with his girlfriend appeared in 1962. Wearing white gloves in the daytime, she must have been a classy type. The artistic style of this advert has stood the test of time well. (Author's collection)

By the latter part of the 1960s, petrol-pump design had changed radically as forecourts switched to self-service. The Wayne 1830 clearly has all the essential features we expect to see when we fill up today – automatic resetting to zero, automatic cut-out and clear dials are all to be found on this apparatus. (Chris Barker collection)

6 PUTTING HIS STAMP ON THE FORECOURT TRADE

The most widespread giveaway to pick up at a petrol station from the late 1950s onwards was undoubtedly Green Shield Stamps.

Introduced in Britain in 1958 by Richard Tompkins, thousands of supermarkets, corner shops and petrol stations offered them to eager customers.

Broadly speaking, when buying petrol or diesel, a motorist would be given five stamps for every gallon. This multiplied if the petrol station was offering double, treble or even quadruple stamps. These would then be stuck in a book to be saved in sufficient quantity to redeem for gifts from either a glossy mail order catalogue or, later on, a chain of Green Shield shops dotted around the country.

By late 1969 Richard Tompkins was the millionaire chairman and managing director of a business empire worth £15 million. He came from a humble background in north London, leaving school at 14 to work for a poster printer, before becoming a laundry boy collecting bag washes in Hornsey during the mid-1930s.

He was interviewed by the AA's *Drive* magazine in late 1969 when he was 50 years old. Tompkins said his life changed for him as a teenager when he read Dale Carnegie's book *How to Win Friends and Influence People*, going on to achieve his ambitions without cutting anybody's throat.

Warning to stamp collectors:
The Triumph Toledo does 33 to the gallon.

You aren't going to fill up your stamp books as quickly when you have a Triumph Toledo. But you won't need to fill up your Toledo as often.

It turns in a very economical 33 mpg. In fact Motor actually got 35·6 (touring).

The Toledo's turning circle is another economic advantage : 29-75 feet. It gains you a lot of ground when parking where every inch you can find counts.

And the Toledo's tight handling and agile manoeuvre also gain you a lot of ground in trammed traffic.

Behind the wheel there's ample room for six-foot drivers to be happy and comfortable all day.

Other keep-you-happy features are : a powerful heater demister with two-speed booster, plus face-level fresh air vents at each end of the fascia. And the controls are as easy to understand as they are to use.

Passengers, too, enjoy a smooth ride on generously proportioned seats. And enjoy luxury touches like full carpeting, arm rests and coat hooks.

It's also a very quiet car. Engine, gearbox and body have been given rigorous treatment to deaden a great deal of noise and vibration.

Suspension is independent at the front ; 4-link 'live' axle system at the rear.

The Toledo may reduce your progress as a stamp collector, but it doesn't hold you back on the road. 0-50 takes 12·5 seconds. 2-door £955·63, 4-door £999·38ex-works inc. p.t. and fitted front seat belts.

Triumph Motor Company Ltd, Coventry. Telephone: 020 3-75511.

🏁 **Triumph**
Triumph put in what the others leave out.

Tompkins said it wasn't until he went to America on a business trip in 1957 that he realised how trading stamps could quickly win new customers and push up sales. Trying to introduce the idea in the UK a year or so later though proved an uphill struggle at first, with his twelve salesmen on the road instructed to phone in every time they won a new customer 'to give the rest of us heart'. The company's big break would come in the early 1960s when Tesco started offering stamps to its shoppers.

By 1969 Tompkins was enjoying his success but by millionaire standards lived modestly. He drove a new Rolls-Royce Silver Cloud bearing the personal plate of 007 but lived in a small flat in central London, having parted from his wife sometime before. He spent his evenings, the magazine said, at the cinema or listening to Andy Williams records.

Some 5 million families across Britain were filling up books of green stamps and redeeming them for 10 million items each year. Typically, a thirty-two-page book would need 1,300 stamps to fill it. 40 per cent of motorists saved stamps, constantly looking for the 8,300 petrol stations and garages that offered them – a quarter of all those in the country. Some would drive miles out of their way to find a place that gave Green Shield Stamps if their local petrol station didn't – and pacify an otherwise angry partner who might be missing out on the chance of a clock radio, a canteen of cutlery or a new bike for the kids.

Green Shield's wide range of quality gifts included housewares, the ubiquitous wine glasses – handy if you hadn't been able to make up a set through other petrol station offers – jewellery, children's toys, gardening, DIY tools and sports equipment.

Inflation in the early 1970s and the fact that so many places were offering multiple rates of stamps led Green Shield to produce the more convenient Big 10 x stamp and a special book in which to stick them. (Author's collection)

The founder and head honcho of Green Shield Stamps, Richard Tompkins, was photographed for AA's *Drive* magazine in 1969 astride a child's go-kart and bearing a candelabra. These were just two of the many gifts available in the company's catalogue as was the teddy bear held aloft by a glamorous assistant. The picture was taken in the car park of Green Shield's offices in Edgware, north London, where one of the trappings of Richard's success could be seen – a Rolls-Royce with the registration 007. (AA *Drive* magazine/Author's collection)

There are probably enough filled Green Shield stamp books for some exotic motoring accessories here! Households all over Britain were madly building up collections to exchange them for anything from saucepans and jewellery to toys and even, for 750 bookfulls, a new car. Green Shield Stamps, as it said inside the books, helped you 'get your own back on the cost of living'. (Author's collection)

By law, trading stamps had to have an alternative monetary value and by 1969 a filled book was worth 8 shillings (40p) in cash – so it made much more sense to redeem a book for the equivalent of 15 shillings (75p) from the range of 1,300 gifts in the catalogue.

'Why did Green Shield Stamps become so popular?' was *Drive*'s inevitable question. Tompkins reply was: 'Cash is ordinary. Nobody saves the money that is left over when they buy anything cut-price. Stamps add excitement, there is the fun of choosing from the catalogue, going along to the gift centre – and perhaps changing your mind when you see the display there. Some people wonder how it is they can get so much for nothing.'

Getting something for nothing was far from reality, of course. Stamps were a means of raising sales and cost money. Green Shield sold stamps to retailers and petrol stations at an average cost of 2.5 per cent of turnover through the till – put more simply 6*d* (2.5p) for every pound.

Petrol stations were able to offer fewer stamps in monetary value compared to a shop because fuel duty on every gallon was factored out of a transaction. This would be more than made up for with the growing number of double or more stamps promotions.

The 1970s saw rampant inflation and petrol prices shot up accordingly – but that meant you could save up those green stamps faster, didn't it? No, the Green Shield company simply raised the number of books you needed before you could get your hands on that shiny Black & Decker drill for Dad or a fake pearl necklace for Mum. Prices went up so much that Green Shield realised saving could become a futile exercise, even with treble or quadruple stamps being routinely offered on many forecourts, and so it allowed customers to claim their gifts with a mix of completed books and cash.

In summer 1973, this move ultimately saw the company morphing into the Argos catalogue showrooms but the stamps continued to be available until 1983. They were reintroduced, for a short time, four years later on a reduced scale but were discontinued altogether in 1991.

If, by some remarkable feat, you were able to fill 750 books – requiring 960,000 individual stamps – you could chop them in for a car! By the late 1960s you could save up for a Ford Escort Mk 1, unsurprisingly only available to special order.

At the time, only one person had managed to achieve this. James Prowting, 48, from Catisfield near Fareham in Hampshire, had, according to *Drive* magazine, taken six years to claim the biggest prize in the Green Shield catalogue.

Mr Prowting, who happened to be a Southampton city councillor, said: 'I thought we would never do it! I drive a 2.4-litre Jaguar, which uses a fair amount of petrol, and the children have two cars. We all agreed to buy fuel only at petrol stations giving quadruple or triple stamps.'

Thousands of petrol stations, large and small, up and down the country gave Green Shield Stamps to customers when buying petrol. In 1964, Stanpit Garage and Village Stores in Mudeford near Christchurch, Dorset, was no exception. The panels of its Cleveland pumps were adorned with large stickers while a banner hung between the end of the building and the Cleveland totem to promote Green Shield. (George Knight)

Mrs Prowting did all her shopping at stores giving away the stamps and it was a moment to savour, said *Drive*, when her husband told the company that he wanted to redeem them all for a car.

'They nearly fell over with surprise,' he said.

After this, the family chose to save for more modest items, including furniture and a motor mower.

This leaflet has been designed to fit into your wallet or purse. Please keep it with you for reference next time you visit your garage.

R. J. A. SERVICE STATION

Canterbury Road East

Ramsgate, Kent CT11 0JX

Telephone: Thanet 57361/2

V.A.T. Registered No. 202 1232 43

Green Shield Trading Stamp Company Limited
Green Shield House, Station Road,
Edgware, Middlesex.

O64/HC/7/72

it has been brought to our attention that some small confusion might exist as to the correct issue of Green Shield stamps –

By 1973, virtually every petrol station was handing out double stamps – but customers didn't know what that meant so Green Shield printed a pocket-sized leaflet to explain. The base rate was five stamps per gallon. A site offering double stamps would give ten per gallon and so on. Stamps were only given to cash customers – those paying by cheque missed out in their quest for new bathroom scales! (Vivienne McIntyre)

7 A LONG GOODBYE TO THE PUMP ATTENDANT

Expecting customers to fill their own fuel tanks was an idea that was initially slow to catch on in Britain.

Petrol stations began switching over in earnest from using attendants from the late 1960s onwards, but credit for the country's first self-service site must go to Turnbull's of Plymouth, Devon, who started the ball rolling in the earlier part of the decade.

Already an established business of more than fifty years standing, it introduced nine self-service pumps to an unsuspecting public at its Breton Side premises on 11 April 1963. This was a full year before self-service began in the USA.

Three of the pumps were Swedish-made and designed to blend five intermediate grades of petrol between Regular and Super, known as Supermix, while the others delivered the standard grades.

Without attendants, staffing costs had been slashed and so Turnbull's could sell petrol at a discount compared to competitors.

Turnbull's would also take the distinction of building the country's first filling station on the outer edge of a roundabout (see front cover photo). It had taken several years to persuade the Ministry of Transport this would not interfere with traffic. Its circular layout featured the nine Shell pumps in the centre of the site with up to seventy-six cars an hour able to fill up and pass around them to get back on to the road.

In later years, the petrol station site became a large stationer's outlet before itself succumbing to the bulldozer.

In Spring 1972 the AA's *Drive* magazine reported on the development of the self-service petrol station, an idea that had initially been slow to catch on in Britain. Thoughtfully, *Drive* produced step-by-step diagrams to show members how to use the new pumps and accompanying coin and pound note acceptor machines – all in the days before credit and debit cards had become widespread. (AA *Drive*/Author's collection)

At the beginning of the 1960s, there were eighteen oil companies operating petrol stations in the UK. By 1972 that number had leapt to fifty, according to the AA's *Drive* magazine – and with that increase came smarter ways of selling more petrol.

It was quickly realised that, apart from reducing staff costs, self-service actually saw customers buy more petrol when filling up.

In 1972, BP's average purchase on an attended site was a little under 4 gallons (about 17 litres), but that rose to 4.2 gallons (19 litres) at a self-service location. Esso rated average sales higher; 4.1 (18.5 litres) with an attendant, to 4.4 gallons (20 litres) when the customer used the pumps. Some 60 per cent of motorists served themselves with more petrol than they had first intended.

With customers prepared to fill up their own cars and then buy about 10 per cent more than they might have done, it was little surprise the operators fell over themselves to convert their sites to self-service.

These new-look sites featured large, well-illuminated canopies to provide some cover, big shops – stacked with a wide range of tempting high-profit-margin items – and, occasionally, super loos.

Eased into a new retail world, customers would eventually forget their resistance to self-service and take full advantage of multiples of Green Shield Stamps along with famous giveaways such as football badges and party tumblers. Stocking up with spare parts for your car, map books or a packet of cigarettes became far easier, too.

Drive magazine reported that National Benzole was selling more of its petrol through self-service stations than any of its rivals, its then managing director Mark Mothio saying he did not believe the concept was being forced upon motorists – it was what they wanted and soaring sales figures were proof of that.

Drive's readers, though, told a somewhat different story. Just 13 per cent of those asked said they preferred self-service and 40 per cent said they avoided the places if they could – two-thirds of them women.

Forecourt operators reported that every time they converted a site to self-service, the gallonage jumped anywhere between 50 and 200 per cent. Customer resistance was being overcome with 1,500 self-service stations already in existence by 1972 and another ten coming on line every week, with BP opening one around every three days.

Sounding a more sombre note, it was predicted that by 1980 the total number of petrol stations would be reduced from 38,000 to 25,000, with half being self-service. As it turned out, self-service sounded the death

By the 1990s petrol station shops had started to take on a more uniform style, as can be seen in this view at a BP site in Milton Keynes. Against the plate-glass windows there are racks of papers and magazines, while sweets are on display in front of the tills with cigarettes behind the cashiers. (© BP plc)

knell for smaller, independent locations and only a handful of attended sites would be left by the end of the 1970s.

For its locations that were expected to stay open through the night, BP had installed 250 £1-note acceptors around the country with another 150 in the pipeline. The late 1960s and early '70s were, of course, a time when you could buy a considerable quantity of fuel for £1 – perhaps as much as 3 gallons (14 litres). Electronic payment cards were still a world away, although credit cards had just arrived on the scene with Barclaycard launching in 1967.

There were seemingly endless teething problems with these machines – some customers being so confused over how to use them they simply got back in their cars and drove off rather than buy any petrol at all.

One Welsh dealer overcame the difficulty on his forecourt by setting up voice-recorded instructions that turned on automatically as a car drew up. BP took to the idea and was planning a version of its own at the time, said *Drive*.

Motorists who preferred self-service said it was quicker than waiting for an attendant – in truth the average time taken to pump a gallon or two was much the same as someone doing it for you. It just seemed quicker because the customer was taking an active role. Factor in waiting for an attendant, who might be checking another customer's oil, and the time saving is more obvious.

Self-service sites across the country were being equipped with pumps that turned on and off as the petrol nozzle was lifted from its holster and selection of either two-, three- or four-star petrol was done at the push of a button or the turn of a dial.

By the end of the 1960s more and more petrol stations were becoming self-service operations with just one member of staff, in a warm booth, needed to keep an eye on all the pumps from a control panel. Here, an employee at Frank Holland Motors, Cambridge, switches on a pump for a customer at the far end of the forecourt. (Chris Barker collection)

From when it was OK to get leggy young women to pose with an inanimate object to make it more eye-catching, comes a 1973 launch photo of a new pump for the self-service era. Designed by Elliott Noyes for Mobil, the cylindrical pump offered greater ease of operation and clearer dials showing both gallons and money. Five-star petrol is sold at 36.5p per gallon, its 100-octane rating matching today's aviation fuel! (Chris Barker collection)

The operators' research showed many men found using the equipment a real challenge – one they nonetheless accepted without first reading the instructions appended to the pumps! In 1972, *The Times* reported that BP staff had seen one hapless motorist scratch his head several times by a new pump, push a pound note up its nozzle and shout, through cupped hands, at the pump: 'Four gallons of commercial, please!'

Regular stupidities on the forecourts included filling up with a lighted cigarette in the mouth, putting the fuel nozzle into the wrong hole and holding a naked flame over the petrol tank to check the fuel level!

More common errors were overlooking the start button on the pumps, forgetting to replace the fuel cap on the car and – about ten times a week at most self-service stations – driving off without paying.

...dern Avery-Hardoll petrol pump in ...w Shell decor, in which the emblem ...s on the pump panel as a replacement ... normal globe.

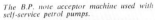

The B.P. note acceptor machine used with self-service petrol pumps.

Being expected to serve yourself was a cultural shift for many in the late 1960s – and not always popular. Not only did the oil companies have to educate people about using new pumps but they also had to show them how note acceptors worked. Here, one customer gets to grips with one of BP's machines. Her single pound note would buy more than three gallons worth of fuel provided the machine was working as it should. (Chris Barker collection)

A slightly battered Hillman Imp makes its way past a Heron operated self-service Texaco site in Christchurch Road, Bournemouth, one day in 1978. The fundamental design of petrol stations had changed dramatically during the decade since the arrival of self service. Forecourts were now covered by overall canopies to keep customers dry as they filled up while nearby, space was set aside for car wash machines – hopefully of the type that didn't harm paintwork. (Ed Coldrick collection)

Opposite: Mobil self-service came to Ramsgate, Kent, in March 1973 when Roy Alexander opened his petrol station on Canterbury Road East. It featured the design-award-winning cylindrical Mobil pumps on the canopied forecourt. Opening offers included 20p off every four gallons – and a chance to win a trip to Calais via Hoverlloyd's hovercraft service, 2 miles away at Pegwell Bay. The site also had a workshop where cars could be serviced and Crypto tuned. Today the site sells used cars. (Vivienne McIntyre)

Parkers' dream

Shop at Shell Motorists' Shops, where the parking's free–and easy!

All you want are one or two things for the car and a packet of fags—but you can't park anywhere. It's a nightmare of traffic wardens, double yellow lines and no entries.

Let Shell dispel that nightmare. There are more than 400 Shell Motorists' Shops up and down the country. They sell everything for your car, as well as soft drinks, sweets, chocolates and cigarettes.

The parking's free – and easy. You'll be safe from the wardens while you buy everything you want from Shell Motorists' Shops.

You can be sure of Shell

With self-service came investment in forecourt shops, as this advert from 1969, with carefully placed models making the place look busy, demonstrates. The range of items was still limited to car parts, sweets and cigarettes. It would be some years before bread, milk and flowers of uncertain quality were added to the stock list. (AA *Drive*/Author's collection)

The Arab–Israeli war of 1973 and consequent price rises in crude oil threatened supplies to the western world. As a result, the British government printed millions of petrol coupons in readiness for rationing. They had been supplied to Post Offices across the country before it was realised they weren't going to be needed. (Author's collection/Rochester Airport Garage Museum)

8 THE GLORY DAYS OF THE PETROL STATION GIVEAWAY

Football badges, wine glasses, socks, tights, soup bowls and, not least, trading stamps of various hues have all been given away on countless forecourts in determined bids to keep motorists loyal to a brand of petrol.

Almost unheard of now, thanks to the need to keep prices down against supermarkets, petrol station giveaways were seen as a guarantee of growing, as well as holding on to, business throughout the 1960s, '70s and '80s.

At a time when more women were learning to drive, and sometimes paying for their own fuel, the 1960s saw pairs of tights – then a relatively new invention in female hosiery – being offered, at a discounted price, with petrol sales for the first time.

In 1966, then little-known actor Christopher Timothy became a pump attendant for a Cleveland TV advert breathlessly telling viewers a pair of tights could be had for the princely sum of 4/11*d* (24p) when purchasing three gallons. *Carry On* actress Hattie Jacques then shows up with a knowing look in her eye to foil the punchline of Timothy's, 'Pulls the birds in, don't it?'

With the Apollo 11 Moon landing of 1969 came Shell's major foray into large-scale giveaways with its set of sixteen Man -n Flight coins. This commemorated the Wright brothers' first take-off to Neil Armstrong making his giant leap for mankind from the lunar module.

A set of Man In Flight coins issued by Shell *c.* 1970 that celebrated sixteen aviation milestones from Icarus and Daedalus of ancient legend to Apollo 11 landing on the Moon in July 1969. (Author's collection)

Football offered a rich seam of interest for Esso's giveaway schemes, its first coming in 1970 for that year's World Cup. England had won the previous tournament four years earlier in 1966, of course, and hopes were high – aren't they always – the team could retain the Jules Rimet trophy in Mexico. A series of thirty coins depicting the entire squad were specially minted and could be inserted into a card with holes cut out for them. Interest did wane though once England had been knocked out!

A year or so later Esso came up with its football club badges collection. A beautifully simple idea that appealed to even the most reluctant fans across the UK, customers would receive a sealed packet containing two self-adhesive badges with every 4 gallons purchased and stick them on a collector card. There were a whopping seventy-six badges to collect and, looking at the near 3ft-wide card, getting the whole set looked daunting. Help was at hand, though, as Esso thoughtfully offered a starter pack of twenty-six badges for the princely sum of 20p. Seeing as we were still getting used to decimal coinage, that was a pricey 4 shillings in old money – but you wouldn't have been able to make up the set without it.

Cardiff City was the last one I needed to finish off my set – never have I been more interested in the principality's leading exponents of the beautiful game – and that was acquired by swopping it for a new table tennis ball with a school chum. Less exciting were the emblems of Preston North End, Bury, Stoke City or Wrexham, which seemed to turn up with annoying regularity. A special journey to the nearest Esso station was wasted and another ten-day wait ensued before my dad would fill up the family Vauxhall Viva again in hope of getting Leeds United or another of the First Division's then big kickers.

Well-known football manager Joe Mercer put his name to a dream team of sixteen of Great Britain's best players in 1971 for Cleveland. His choices included Bobby Moore, Billy Bremner, George Best, Gordon Banks, Colin Bell and Geoff Hurst. (Author's collection)

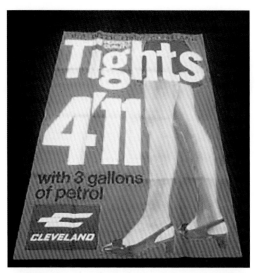

Clockwise from left: A 5ft-deep poster as used on the forecourt totem advertising Cleveland's 1966 tights offer – 4/11*d* (24p) when buying 3 gallons of petrol. (Author's collection); Shell's giveaways tended to steer away from the ever-popular football subjects, as this 1970 set of twenty Historic Cars coins demonstrates. Coins in this series ranged from the 1886 Benz three-wheeler to the hot and happening 1970 Lamborghini Miura. Descriptive notes on the back of the card were by leading motoring historian Michael Sedgwick. (Author's collection); The centenary of the FA Cup in 1972 prompted Esso to produce a set of coins depicting the badges of all thirty teams that had previously won the trophy. A sumptuous collector card, which had a fully illustrated booklet wrapped around it, there was additional space for a gold coin naming that year's winner. Available just days after the cup final, it was Leeds United who took the honour. (Author's collection)

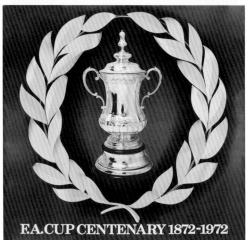

Arguably the daddy of all the football-related petrol-station giveaways at the time was Esso's Football Club Badges collection of 1971. There were seventy-six badges in total to collect and stick on a 3ft-wide card – but a starter pack of twenty-six badges, sold for 20 new pence, helped to make the task easier. Those twenty-six were only available in the pack so you Couldn't complete the set without it. (Author's collection)

Laughingly, the collector card, given free at Esso stations, advises people to keep their badges safe as they 'may become a valuable collectors' item' – today you can buy a full set on eBay for less than £20.

The marketing agency devising the scheme for Esso paid the seventy-six clubs featured just £100 each for the rights to reproduce their logos – we can only guess how much would be needed now in today's far more commercialised football rights arena to repeat that idea.

The centenary of the FA Cup was celebrated in spring 1972 by Esso with a set of thirty silver-coloured coins of the logos of clubs that had won the revered trophy since 1872. The winner of that year's final, Leeds United (who beat Arsenal 1-0 with a rare header by striker Alan Clarke in the second half at Wembley), saw its emblem pressed onto a special gold coin offered just a few days after the event – disappointingly, the only time Leeds has won the cup!

Mobil customers were able to collect tokens for soup bowls in 1973. One was given with every 2 gallons purchased and nine were needed before claiming a dark brown bowl that had a handle that tended to snap off far too easily. Saving for a set of six could take many weeks and called for 1,800 miles of steady driving at 35mpg. Now, that could take longer and about 1,000 miles further with today's far less thirsty cars.

Such was the explosion in freebies by the early 1970s, when every petrol company was handing out stuff left, right and centre it seemed, BBC's *Monty Python's Flying Circus* comedy programme parodied the concept where an undertaker's radio advert promised 'a free set of wine glasses with every certified stiff'!

While all these goodies were being bandied around to win custom, somebody had to pay for it all – and ultimately that wasn't going to be the big oil companies.

The poor old petrol retailer had to foot the bill and it would be considered bad form if they didn't sign up for the promotions, as TV's consumer reporter John Stapleton revealed in a *Watchdog*-style investigation in 1972. Several petrol station franchisees told him they would much rather simply reduce pump prices than have to buy in badges, coins, glasses and all the other trinkets required. They went along with them, mindful they could adversely affect their chances of renewing their site leases if they didn't. They reckoned they could reduce pump prices by as much as 10 per cent instead of suffering infernal giveaway offers. In those days a gallon of petrol cost around 33p – 6/8*d*

in old money – so a 3p a gallon discount would have been worth having. Oil company bosses countered they were merely providing what the consumer wanted and the offers had proved a great success for all involved.

As it turned out, promotions suffered in late 1973 amid an oil embargo imposed by the Middle East exporting countries during a war with Israel. The pumps in the West quickly started to run dry and the British Government decided to issue tokens of its own – otherwise described as petrol ration coupons. Happily,

Perhaps Esso was milking the football theme by the time it came up with its Top Team series of twenty-two soccer superstars in 1973. These were small metal picture discs that needed strong tape or glue to hold them to the card. (Author's collection.)

they weren't needed but there were long queues at the pumps and prices went sky high – suddenly, discounting, however small, became the in-thing as people quickly started to tire of giveaways.

Offers for free sets of glasses came and went in the 1980s and '90s, but it was clear that giveaways had had their day. With supermarkets opening out-of-town superstores and petrol stations in the late 1990s came loyalty cards and the ability to pick up points on the whole week's shopping as well as petrol or diesel.

Giveaways enjoyed a small resurgence in 1993 when the puppet-centric *Thunderbirds* TV programme made a comeback on to our screens and a set of coins depicting its main characters was offered by Fina. It was a poor substitute though for the model of Tracy Island that had taken the shops by storm that Christmas.

Nearer the end of the decade, Sainsbury's did its best to induce some World Cup fever with a set of coins depicting the England squad. This also helped the supermarket build its then fledgling share of the petrol business.

Since then, keeping down pump prices has never been more vital and the days of ripping open a small envelope in hope of acquiring a Chelsea, Liverpool or Celtic football emblem are now long gone.

Hopes were high in 1970 that England would retain the World Cup in Mexico and accordingly, Esso brought out one of the first series of giveaway coins with its collection showing the thirty-strong England squad. Subjects included Alan Ball, Alan Mullery, Jack Charlton, Norman Hunter and Francis Lee. (Author's collection)

If you were struggling to complete your set of Esso Football Club Badges, the oil company thoughtfully provided a service enabling you to fill the empty spaces on your card once the promotion had finished. You had to order at least five badges for the princely sum of 10p. If you needed more, it was 8p for every five badges thereafter. (Author's collection)

You would have to drive between 350 and 400 miles to burn up the petrol to get you enough tokens for just one soup bowl from Mobil when this promotion was running in the summer of 1973. By now, two-car households were becoming more common so collecting tokens was a little easier – provided your other half remembered to go to a Mobil garage as well – so you stood an outside chance of having a set of six bowls. Chicken and mushroom anybody? (Vivienne McIntyre)

Besides collecting Esso's Top Team picture discs, there was a chance to win a holiday for two to the 1974 World Cup tournament in West Germany. Scotland, Wales and Northern Ireland fans would have been delighted to go as England failed to qualify, after an historic defeat to Poland! (Author's collection)

Cleveland opted for a less mainstream topic when it gave away small coins forming its Historic Campaign Medals series *c.* 1973. You didn't get the full effect of the subject unless you bought the collector card upon which to stick them. The ribbons printed upon it set off the medals well and brought much-needed colour. (Author's collection)

Thunderbirds were definitely go for Fina in 1993 when its petrol stations offered coins from the TV series, which was making a major comeback that year. Arguably only slightly more popular than the coins were copies of *Blue Peter*'s instructions to make the model of the Thunderbirds' Tracy Island base. (Author's collection)

Mobil tawny glass offer poster from 1973. (Vivienne McIntyre)

FUELLING UP IN THE URBAN JUNGLE

Showing all the hallmarks of a posed promotional photo of the 1960s – although the men in the centre look like they might be about to have an argument over a small wager – the service station on Cwmbach Road, Aberdare, would have been built to BP's corporate design of the era. Its Super Plus petrol is on sale at 3/9*d* (19p) per gallon according to the pump sticker. (Ed Coldrick collection)

A petrol station without any petrol pumps was a novel idea in the 1960s when BP opened for business in the Birmingham suburb of Acocks Green and was claimed to be the first in Europe. Customers would serve themselves by pulling down one of the electronically operated hoses set in the roof of the canopy. Although innovative for the time, it was one of those ideas that didn't catch on with the paying public. (Chris Barker collection)

Parkhurst service station in West Molesey, Surrey, was a place that represented a new approach for BP sites in 1966, being developed with a specific car parts shop. A customer does some earnest window shopping while the forecourt is clear. A diesel pump takes its place on the nearest island, perhaps marking an early shift from siting them away from everything else? (© BP plc)

Canopied forecourts with their multi-hosed pump islands, an automatic car wash and a well-stocked shop and pay area were all established features of modern forecourts by the end of the 1980s and early 1990s when this image was captured in Earley, a suburb of Reading. (© BP plc)

Opposite: In the 1960s many people, not just Mods, bought mopeds to go shopping and meet friends – hence this image of a pearl-necklaced Essex woman having her 175cc Lambretta refuelled by a BP attendant in Rivenhall near Witham. Riders realised the oily bits of a moped were covered up so clothing was more likely to stay clean en route. The Avery Hardoll 'Petroiler' dispensed blends of petrol and two-stroke oil by the pint and could be wheeled to anywhere on the forecourt. (© BP plc)

An attendant sets to work filling up a station wagon – a van with windows – while its driver fumbles around for payment at BP's petrol station at Thornwood, Epping, in 1960. (© BP plc)

This is one of those might-have-been ideas in petrol retailing that didn't work out. Trying to increase a forecourt's throughput of cars, BP installed a rotating pump at its petrol station in Datchet Green, Surrey, in the mid-1960s. Three cars could gather around the pump waiting for the attendant to get to them without blocking each other's entry and exit. The image shows the rotating arm clearly as well as a cabinet for holding cans of oil and maps for sale. (© BP plc)

The solus forecourt concept of one brand of petrol at all pumps hadn't reached Falles garage at Bagot Road in St Saviour, Jersey, when this picture was taken in the mid-1960s. We see a pair of twin-grade Esso and BP pumps at either end of the island with oil can racks in the centre. (Ed Coldrick collection)

Three-star economy petrol is getting an unusual boost here via a poster fixed to the totem sign at A&R Thomas garage in Bedford Road, Kempston, during the early 1970s. Twin-grade pumps, fitted with long swing arms, await customers, while further back there is a separate Pink Paraffin pump. (Chris Barker collection)

Not a public petrol station but a privately operated fuel island owned by Manchester-based haulage company Edward Beck & Son Ltd at its depot in Greg Street, Strockport, in 1967. Note the 'not for resale' globes on top of the pumps and the higher-than-average canopy enabling its fleet of lorries to park up close. The company previously held the contract to build Shell stations in north-west England. (Teddy Beck via Edward Beck history website)

A 2,000-gallon Dodge petrol tanker, operated by Isherwood's of Eccles, makes a delivery of VIP fuel to Cockshoots garage in Manchester, sometime in 1962. VIP's 95-octane petrol was selling for 4/6d (22.5p) per gallon at the time according to the sticker on the right-hand Avery Hardoll pump. You could save 5d (2p) per gallon by opting for the 83-octane product on the left-hand end. Meanwhile, discounts of £35 were being offered on the used cars behind. (James Kay via Paul Anderson)

During the late 1950s and early '60s Contractors (Manchester) Ltd, which would become Edward Beck & Son Ltd, were contracted to build all Shell petrol stations in the north-west of England. The company would level the site, install the underground storage tanks and build the entire forecourt. This 1961 picture shows construction of the Shell site on the A6 London Road, Hazel Grove near Stockport. (Teddy Beck via Edward Beck history website)

The Douglas Garage of Sheep Street, Northampton, had one of the first Triumph Heralds in its showroom when photographed in 1960. Judging by the relatively tucked-away position of the BP and National pumps, selling petrol wasn't big business for the garage but a Standard 10 appears undeterred. (Author's collection)

A few miles west of Manchester lies the suburb of Flixton, where this petrol station, Flixton Autopoint, was pictured in 1985. In those days, it was a Burmah retailer. Unleaded petrol being sold by the litre was still yet to happen judging by the signage. The forecourt still trades but under the Texaco banner. (Chris Barker collection)

The Roman Catholic church of Our Lady of Fatima was still under construction in First Avenue, Harlow, Essex, in summer 1957 and contrasts with the busy scene at Kenning's petrol station in the foreground. A Shell tanker is making a delivery and customers are being served from the Avery Hardoll pumps. On sale in a cabinet are cans of Shell X-100 oil. (Photographic services, Shell International)

Art deco-style came to Walmer, a village just to the south of Deal in east Kent, when Henly's took over what had been County Motors on Dover Road. Pictured in the 1960s, when it was still possible to buy a second-hand Austin Cambridge for £420, the site was later taken over by Adamsons Motors, who demolished the showroom in 2012 and built an Esso petrol station in its place. (Author's collection)

Hungerford Filling Station stood close to the centre of town in Charnham Street in 1959 when it sold Mobil's range of petrols and bottles of oil. (Hungerford Virtual Museum)

A gracefully ageing RT London bus trundles past Keelers Motors of Harrow on the Hill, on its way to Golders Green Station, sometime in the early 1970s. Car and van hire was a big part of the Keeler business but there was still money to be made from petrol sales – as the Skoda and Ford Escort testify as they line up by the Shell-branded pumps. (Chris Barker collection)

Photographed in 1985, the Jet petrol station in Leopold Street, Birmingham, later became a Burmah site but now stands empty awaiting redevelopment. (Chris Barker collection)

A good example of the urban backstreet garage formed of old buildings and corrugated iron sheeting – where who knows what nefarious dealings went on – is that of Manor Service and Filling Station in Wheeler Street, part of Birmingham's Lozells district. This comes from a photograph taken in June 1960 by surveyors acting for a local transport company. The location today is entirely different. (Chris Barker collection)

The funky Markham Moor petrol station in Retford, Nottinghamshire, was built in 1961 for the Lincolnshire Motor Company beside the southbound carriageway of the A1, at the junction with the A57. Costing £4,500 to build as a hyperbolic paraboloid, it was designed by local architect Hugh 'Sam' Segar Scorer with Hungarian structural engineer Dr Hajnal-Konyi. In 2003 plans were unveiled to demolish the canopy but the Highways Agency reconsidered after protests. The structure has lain derelict since March 2012 when it was Grade 2 listed. (Chris Barker collection)

Built in the early 1930s opposite Sutton Manor Colliery, in St Helens, Lancashire, is Parkside Garage, seen here in 1959. It was opened by William Hancock, who started a haulage business in 1926 after winning £500 in a newspaper football pools competition. His wife, Isabel, predicted the vital twelfth result — a game involving Aston Villa — after he was uncertain when filling in the coupon. They built a house close by and named it after the team. The petrol station still exists. (David Hancock)

Forming part of a parade of shops in Birmingham's Selly Oak *c.* 1960 was the forecourt of Pershore Road Garage, which boasted a four-strong line-up of National- and Power-branded pumps. Cars pulling up at the kerb for petrol would have been reached by fuel pipes on overhead swing arms. (Andy Maxam, maxamcards.co.uk)

How convenient this Morayshire-registered car bearing the local SO suffix was available for a photograph at Proctor & Paterson's Esso petrol station in Elgin sometime in the early 1960s. A cheery assistant wipes the Austin A55's windscreen while behind her is a rack of glass bottles of oil. It's not clear if the car, and the registration, belonged to Esso but as a cherished number today it would be worth several thousand pounds. (Ed Coldrick collection)

BP's Refinery Service Station, on the doorstep of its Grangemouth oil refinery, began operations in March 1971. Director Alan Roberston performed the official opening, followed by a sumptuous lunch for guests in the refinery's information centre. The petrol station was equipped with four 'post-payment' self-service pumps, a manned lubrication bay, automatic car wash and a BP shop. (Chris Barker collection)

Rhiwbina, a northern suburb of Cardiff, was home to Cliff Smart's garage business during the late 1950s, seen here with a row of fine used cars, waiting to become prized classics! Three Morris Minors are visible along with a Vauxhall Wyvern and an MG Magnette. The Shell pumps look impressive as they await customers. (Ed Coldrick collection)

Assuming the film *Fast Lady*, as seen on the local Odeon poster boards, was newly released at the time, then this photo was taken in early 1963. Leonard North's garage and petrol station was located in Surrey's Thames Ditton High Street. Not far away was the headquarters and factory of AC Cars, which may go some way to explaining why there is an example in the doorway of these premises. The pumps show Shell petrol is on sale for 4/4*d* (22p) per gallon. (Ed Coldrick collection)

This flying-saucer-styled petrol station was the Tower Garage, found in Alderley Edge, 15 miles south of Manchester, in 1963 when Total supplied the petrol. (Chris Barker collection)

BMC agent Hoggar's Garage, of Victoria Road East, Leicester, boasted a trio of tall Shell globe-topped pumps in the late 1960s. A two-stroke blending machine stands further away while a customer peers through the shuttered main doorway, possibly seeking assistance. (Terry Dand)

The market town of Leek, Staffordshire, was where Ferdinand Bode set up his bicycle repair business around 1910. Half a century later, when this image was taken, the firm was well established at its Buxton Road premises, selling and servicing cars including Standard Triumph, Jaguar and Vauxhall. A trio of Cleveland pumps ensured petrol was readily available for them. Cars are still serviced here but the front of the building is a separate shop these days. (Neil Collingwood)

FUELLING UP IN THE RURAL BACKWATER

In a 1950s or '60s oasis of tranquillity beside one of Scotland's main highland roads at Nether Lochaber, Onich, Angus Cameron stands in his store's doorway ready to supply travellers with all they need for their journey – including Shell petrol. Meanwhile, Mrs Cameron appears outside their adjoining home to say lunch is on the kitchen table and he had better step round smartly before it goes cold! Today, the Old Store, as it is now known, forms holiday accommodation. (Chris Barker collection)

Charing Motors, found on the A20 about a mile from the centre of Charing, near Ashford, Kent, has been run by brothers Steve and Dawson Blake since they took it over from their father, Ralph, in 1981. He bought the garage in 1969 after working for another garage nearby. One of Britain's first Happy Eater restaurants opened on the end of the site in the early 1970s but has since been replaced by a modern vehicle workshop. (Author's collection and the Blake brothers)

Today Charnock Richard is just another motorway service area on the Lancashire stretch of the M6 – in fact it was the first to open on the highway. Back in the mid-1950s, three Fina pumps, attended by respectable ladies, stood outside Smith's garage and transport cafe on the Preston Road towards Chorley to meet the fuelling needs of passing traffic. (Ed Coldrick collection)

What looks like a former London taxi has found its way out of the capital to the quieter climes of Stowmarket, Suffolk, in 1970 where it rests at Combs Lane Filling Station. Serving BP petrol, diesel and two-stroke blends, the site also gives Green Shield Stamps to its customers. (Chris Barker collection)

Tall pumps stand sentinel outside Edwards Garage at Dudleston Heath, Ellesmere in Shropshire, around 1930. The staff seem reticent to pose for a photo but have paused to gather round all the same. Signs for petrol and oil on the building itself add period charm. (Ed Coldrick collection)

A mobile shop vehicle stands at one end of the forecourt of Frating Garage to ply its trade to the customers of the Essex village of Elmstead Market near Brightlingsea, c. 1970. The garage, set on the B1029, dates back to the early 1950s when it was started by Alan Gray and his sister Lillian. A Morris Oxford and an Austin Cambridge are among the cars lined up for sale in the showroom. (Jan Al-Karaghouli)

Gainsborough petrol stations were most likely to be seen around East Anglia from the early 1960s to the early '70s before being bought up by Total in 1974, along with another regional brand, ARCO. This installation is typical of the Gainsborough garages to be found in Norfolk and Suffolk. Pink Paraffin is on sale at this unknown location, too. (David Hicks)

Motoring holidays in Scotland have long been popular and became more attractive to the masses as they bought cars in vast numbers throughout the immediate post-war era. The need for comfort breaks and more petrol was addressed by many enterprising cafes and tearooms on the Highland roads, including that at Glen Shiel, en route to the Isle of Skye, seen here in the early 1960s. (Chris Barker collection)

Ready to man – or woman – the pumps are the staff of Golding & Woolgar Ltd in the small jam-making town of Histon, Cambridgeshire, sometime in 1960. While the female attendant leans against one of the National-branded pumps, her colleague has a hand on the two-stroke blending machine, usually used to help fuel motorbikes and scooters. Window stickers indicate the full range of Eveready batteries and torches is on sale here as well. (Mike Petty collection)

It's common today for a forecourt to boast a sweetshop beside it – but in the early 1950s, it was rare for a sweetshop to have petrol pumps out front. S.G. Crisp, confectioner and tobacconist of Guildford Road, Lightwater in Surrey, was a Cleveland supplier whose pumps feature an array of glass globes from different eras. Another has become part of a totem sign to indicate whether or not the place is open. Today, Guildford Road is home to a BP station and M&S shop. (Chris Barker collection)

In March 1966 Jack Acott expanded his garage on the edge of RAF Manston near Ramsgate, Kent, which included a handsome line-up of Cleveland pumps. Jack Acott began servicing VW Beetles during the early 1950s for American airmen at Manston – between 1950 and 1958 Manston was a USAAF base – who had bought their cars through their on site Base Exchange (BX) store. He became the local VW dealer and sold hundreds more. Today, it's a derelict pile of rubble at the very corner of an abandoned airport. (Author's collection)

Set close to the Norfolk Broads is the pristine garage and forecourt of Landamore's at Church Road, Hoveton. The company has been building boats and motor yachts since 1923 with its fourth-generation management continuing the work elsewhere in the village. This photo was taken in June 1964 when servicing and testing of cars, as well as selling Mobil petrol, provided useful additional income. The building still exists but with no motoring connections. (Archant/EDP Library)

Outside Butler's store in Helmsley, North Yorkshire, in 1984 is this kerbside line of three Avery Hardoll pumps bearing the name of J.R. Rix, a Hull-based family company with interests not only in oil and petrol supply but haulage, shipping and stevedoring. The easing of wartime petrol regulations encouraged John Robert Rix to establish several petrol stations around the Hull area and later to other parts of the region. The prices on these pumps are per gallon, not per litre. (Tim Moore)

Lapworth Garage, in the Warwickshire village of Lapworth and seen here in the 1930s, still exists, albeit a short distance from its original location, which it had occupied since the business was formed in 1916. A young attendant stands by the furthest of the sturdy pumps awaiting a customer, while the man beside the car is just happy for a chance to show off his pride and joy. (Chris Barker collection)

The National, BP and Shell petrol station at Marlesford, Suffolk, appears typical of a rural forecourt in the early 1960s and may have survived largely untouched for many years after. A glass-sided kiosk stands in the centre of the four-strong pump line to shelter the attendant and secure the all-important till, while beyond lies the garage. (Ed Coldrick collection)

The petrol station in the Norfolk village of Melton Constable, was stuck in a time warp even when photographed in September 1969. The tall hand-cranked pumps, topped with Shell globes, would have been about 40 years old then. The corrugated iron frontages seem typical of some old garages and, in their own way, exude rustic charm. Both buildings are long gone but the one further away still stands, albeit unoccupied. (Archant/EDP Library)

Pictured during the early 1950s before solus arrangements insisted that forecourts sell only one brand of petrol, this at Piercy End in North Yorkshire's Kirbymoorside offered National, Dominion and Shell petrols from its line of kerbside pumps. Ryedale's was formed in 1946 and still trades today after much expansion. It continues to hold the Vauxhall franchise for the area. (Chris Barker collection)

Turn right for Warminster or continue straight on to Trowbridge and Bath implore the signs either side of the ironmongers facing S. Hillman's garage and petrol station in Bath Road, Beckington, just a year or two before war would be declared against Germany for the second time. (Chris Barker collection)

In the 1930s the hamlet of Baldwin's Gate, south-west of Stoke-on-Trent, was the home of Whitmore's filling station. The forecourt was one empty expanse as the pumps were tucked behind the glass frontage, where an attendant awaited. Providing cover this way would have been a far-sighted idea at the time. There is no sign of the filling station now, the vacant space providing parking for the Sheet Anchor pub just beyond. (Neil Collingwood)

The triangular tail fin on the kiosk roof adds to the charm of this 1960 photo of the Airfield Filling Station on Yarmouth Road, between Caister-on-Sea and Great Yarmouth, Norfolk. Trollies of bottled oil can be seen by one of the pumps as the attendant fills up a Ford Prefect. Beyond, the aircraft on the right is a 1944 Auster 5, originally built for the RAF but then sold off and reregistered in 1948. It was still flying twenty years later. (Archant/EDP Library)

9 DELIVERING THE JUICE

Moving fuel from refineries to forecourts is a major logistical operation involving hundreds of tankers, miles of underground pipelines, rigorous safety processes and multi-million-pound investments in plant and machinery.

The supply chain is so precise that any prolonged break could see pumps run dry – and the nation's economy come to a grinding halt in a few days.

In the early years of the twentieth century, cans of petrol were moved around the country by rail for onward delivery to forecourts by lorry and even horse-drawn wagons. Later, fleets of oil-company-liveried railway tankers were built to transport bulk quantities of petrol and diesel from refineries to regional terminals from where road tankers would set out to make their deliveries.

Today, while virtually all of Britain's bulk petrol and diesel supplies are transferred via a network of underground pipelines connecting refineries to terminals, it is still the road tanker that makes the final link to the forecourt and is therefore a familiar sight among motorists up and down the country.

GETTING FUEL TO THE FORECOURTS

Clockwise from top left: Amoco is a brand of petrol that hasn't been seen in Britain for some years – it belongs to BP these days – but was a common sight in the 1970s. Here, an AEC Mammoth tractor unit and tanker take fuel to the forecourts. (Chris Barker collection); A three-axle Seddon Atkinson Strato with twin-axle trailer looks resplendent in BP's then new livery as it makes a delivery. (Chris Barker collection); A grey day and a grimy location for a kerbside delivery by one of Esso's smaller vehicles in the 1950s. (Ed Coldrick collection)

Brew Brothers Ltd service station on London's Old Brompton Road was the destination for this imposing Scammell-hauled unit one day in 1954. (Chris Barker collection)

Clockwise from top left: Davis's Garage at Temple End, High Wycombe, Buckinghamshire, is the setting as this National-liveried AEC tanker makes a delivery in the mid-1950s. (Ed Coldrick collection); Leyland trucks were being used by Gulf Oil to transport its products around the country when this image was captured in the 1960s. (Tony Martin); Shell and BP had a joint distribution arrangement across Britain for many years, as illustrated by the yellow-and-white-liveried tanker parked outside Gavson's petrol station in Manchester's Hulme district in 1963. (Ed Coldrick collection)

Opposite: Isherwood's Petrol Company, the business behind VIP Petrol, operated this Foden eight-wheeler unit from its base in Eccles, Manchester. The vehicle was registered in 1964 and could carry 4,000 gallons. (James Kay via Paul Anderson)

Some of the first petrol stations in Britain would have received deliveries from vehicles like this articulated AEC Matador and tanker dating from the 1920s. Thanks to fluctuating Anglo-Soviet relations at the time, Russian Oil Products sometimes proved a controversial brand to sell. (Author's collection)

A pair of Isherwood's-owned tankers fill up at Petrofina's depot at Preston in 1958. To the left is an AEC Mammoth Major eight-wheeler while in the foreground is a Foden eight-wheeler, each built to carry 4,000 gallons. Soon after this photograph was taken, the AEC was repainted in VIP Petrol livery. In 1974 the VIP name disappeared as Elf Petroleum acquired the company. (James Kay via Paul Anderson)

An example of a supermarket petrol tanker, this time in Sainsbury's colours heading along London's Edgware Road in 2016. (domdomegg via Wikimedia Commons CC 4.0)

Some 2,500 petrol-tanker drivers at fuel depots in east London went on strike during October 1953 campaigning for a pay rise and protesting against non-union labour. Pumps started to run dry in the capital and parts of south-east England until 6,000 servicemen were drafted in. This Esso delivery to a forecourt in Rainham, Essex, sees an Army sergeant connecting the tanker hose to the underground tank while his corporal, standing on the tanker, is ready with a dipstick. (Chris Barker collection)

A National Benzole driver carefully checks the levels with the dipstick on his Thornycroft Sturdy tanker sometime during the 1930s. (Ed Coldrick collection)

The modern petrol tanker can carry around 9,400 gallons and those with supermarket liveries have become a common sight on Britain's roads since the mid-1990s. This Tesco Momentum-liveried DAF tanker was seen at Sudbury, Suffolk, in 2013. (Rodney Smith)

10   So What are we Putting in our Fuel Tanks?

Bitter wars have been fought to control the stuff, whole countries driven near to bankruptcy for the lack of it, while other nations have so much of it they can't spend the fantastic profits quickly enough.

Make no mistake, oil is very big business indeed and billions have been invested over the past 150 years in finding more of it to meet the globe's never-ending thirst for this 'black gold'.

Without resorting to an explanation that calls for some sort of chemistry or geology degree to understand, the crude oil from which petrol is refined or distilled is extracted from deep underground, or far below the sea, as a yellowish-black liquid.

In simple terms it's the squashed remains of many billions of zooplankton and algae that died millions of years ago. This fossilised liquid is extracted mainly by huge drilling rigs and pumped through miles of pipelines, some stretching across whole countries, to a network of specialist sea ports and then shipped in vast tankers to refineries across the globe.

Containing a wide range of hydrocarbons, crude oil is turned into petrol, kerosene, asphalt and chemical components to make plastics, pesticides and medicines. It's estimated that, worldwide, we use about 95 million barrels of this cloying gunge every day.

WITH AN EYE TO THE FUTURE

Motorists who remember the excellent quality and consistent reliability of pre-war CLEVELAND petrols keenly look forward to the return of branded motor spirit.

And when that day comes the younger generation will find a new pleasure in the smooth running and additional power which CLEVELAND petrols assure.

XXXV

With an eye to the future

Wimbledon · June 26th to July 8th

CLEVELAND PETROLS LATER

20

By the time we drive up to a petrol-station forecourt and get our hands around a pump nozzle, a range of additives will have been mixed into the finished fuel, which have been designed to help make our vehicle's engines run more cleanly and efficiently.

Early petrol varied considerably in quality and most owners would stick with a brand that suited their cars best if they could. Many engines suffered the vagaries of low-grade petrol coursing through their cylinders and wearing them out prematurely or simply not running at their best.

Experiments in Britain, America and elsewhere found that adding benzole eased the problem to some degree. Many benzole extraction plants had been set up during the First World War alongside collieries, gasworks, steelworks and other industrial centres to extract it from their coal supplies to make explosives and ammunition.

Once the war was over, benzole would be mixed with petrol. This gave rise to the formation of the National Benzole Company in Britain as a co-operative selling organisation by its producers. With a growing number of cars taking to the roads, it eventually moved into selling its own brand of petrol/benzole mix, the company purchasing its petrol from BP. Ultimately, National, as it was known to millions, became a subsidiary of BP and this explains why you would often see the two brands alongside each other on many forecourts of the period.

Alcohol was another ingredient tried by some petrol companies to give their products a bit of a boost, most notably by Cleveland during the 1950s, which claimed it made cold starting far easier. Certainly, owners of the era's wheezy Austins or Morrises were likely to need all the help they could get on an icy winter's morning!

THEY WERE BUSY PRE-WAR
THEY WILL BE AGAIN...

CLEVELAND petrols, always associated with reliability and better performance, have long enjoyed the confidence of discriminating motorists. When individual brands return, the choice of Cleveland will once more be justified.

"The influence of Ethyl has assisted towards improving the breed of British Small Cars."

SIR MALCOLM CAMPBELL

"And Sir Malcolm Campbell is right" said the Designing-Engineer

Those small cars shown leaving the factory for their first journey in life might have had engines of twice the weight or half the horse-power, but for the high-compression fuels in the development of which the proprietors of Esso Ethyl have played such an important part. Modern small cars owe much of their brilliant performance to Esso Ethyl which, by stopping pinking, both audible and *muffled*, makes them livelier, faster and better hill-climbers.

ESSO ETHYL— THE PETROL THAT STOPS MUFFLED PINKING

ESSO ETHYL

THE NEAREST THING TO FLYING

ESSO ETHYL

LOOK FOR THE NEW OVAL GLOBE YOUR GUARANTEE OF QUALITY

Try the two-tankful test

First, fill up with Shell with I.C.A. Do not expect an immediate improvement, but second, fill up with Shell again. This will enhance to work on the deposits already in your cylinders

The invitation a million people accepted

REMEMBER THIS INVITATION? We put it into our advertisements nineteen months ago, when we had just introduced our new additive I.C.A into Shell.

A million people accepted it.

The result? A million people are convinced that Shell with I.C.A is the most completely satisfactory petrol they have ever used.

If you're not one of the million, we give you the invitation again: try the two-tankful test. You will be doing it on Summer Shell, which is now in the pumps. Summer Shell is specially blended to meet the special demands of hot weather and heavy loads—and it contains I.C.A of course. On your second tankful of it you will notice definitely smoother running and fuller power.

THE MOST POWERFUL PETROL YOU CAN BUY

SHELL WITH I.C.A

THAT'S SHELL   ~THAT WAS!

Ask for the new Shell and BP Road Maps (6 miles to the inch) at any Shell and BP garage. Price 1/-.

In fact, the alcohol label was perhaps a little misleading as the additive is better recognised today as ethanol – which now makes up at least 5 per cent, set to rise to 10 per cent, of the content of modern unleaded petrols.

But it was back in the 1920s when the real answer was found to stop engines knocking or 'pinking'. This is the irritating noise a car makes when the ignition is too far advanced (a spark being delivered too early) or the compression inside a cylinder ignites the petrol before the spark is applied. This causes a metallic pinking sound and eventually damages the fuel inlets.

Tetraethyl lead was added to petrol for the first time in 1921 after American inventor Thomas Midgely, then working for the mighty General Motors, discovered the mixture would stop engines knocking – and it would be a key ingredient, as well as a pollutant, for more than seventy years.

Even in the 1920s, there were grave misgivings among scientists over the use of leaded petrol. German chemist Charles Klaus wrote to Midgely telling him: 'It's a creeping and malicious poison', and warned that lead had already killed another scientist.

Ironically, Midgley was taken ill with lead poisoning while helping to plan the production of tetraethyl-fortified petrol, but he still felt able to tell an oil industry engineer that widespread poisoning was 'almost impossible, as no one will repeatedly get their hands covered in gasoline containing lead'.

He wouldn't have known it at the time – he died in 1944 – but he also came up with another of the world's most polluting processes – that of adding chlorofluorocarbons to refrigeration equipment.

Towards the end of the 1920s, Pratt's, later to become Esso, introduced its 'ethyl' petrol and this idea was adopted by several other oil companies in the decade that followed.

Over time, they realised that adding lead to fuel significantly improved the octane rating of petrol. This enabled them to produce cheaper grades of petrol yet still retain the required octane level for a car's engine to run smoothly.

Octane ratings gained popularity in the 1930s as a measure of how resistant to knocking a particular blend of petrol might be. For many years, the oil companies didn't talk too much about what octane

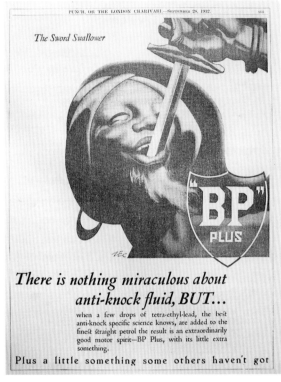

ratings were for. Instead, they introduced 'super' brands for high-performance engines (which have higher compression and are more prone to knocking). Pumps would be marked 'super', 'high grade' or 'ethyl' for this higher-rated petrol and 'regular' or 'standard' for the lower type. The price difference between them was about 1d (0.5p) per gallon.

During the Second World War and in the years that followed all of this went out of the window to be replaced by a single-octane pool petrol. Rated at around a miserly sounding 85 octane, it was meant to be sufficient for the majority of cars still running on Britain's roads.

Once rationing was out of the way and solus forecourt arrangements had taken hold in the early 1950s, Shell introduced ICA – Ignition Control Additive – into its product range and the three red squares bearing those initials on the pumps became a familiar sight from 1954 onwards. In reality, ICA was nothing more than tetraethyl lead.

By the 1960s, with growing calls for consistent quality and car engines becoming more sophisticated, British standards were introduced to ensure compliance and heralded the arrival of the star system in 1967. Forecourts were now able to provide a selection of petrols to suit different-sized engines. Smaller vehicles such as a Mini or a Ford Anglia might be happy running on two or three star while a bigger Ford Cortina, Austin Cambridge or Rover 2000 would need four star. The larger-engined motors of 2.5 litres or more also had the option of filling up with five star at some petrol stations. One-star petrol has never appeared in Britain – and

was probably only best for lawnmowers if it had been – while five star, rated at 100 or 101 octane, fell from popularity after big price increases during the mid-1970s. Today, it is sold as fuel for aeroplanes.

CHARIVARI.—July 13, 1932.

"How on earth did she do it on 2 gallons?"

"That's the Benzole, my boy!"

It is an acknowledged fact that the power of a motor spirit depends upon the amount of aromatic hydrocarbons it contains. It is also an acknowledged fact that Benzole contains a higher proportion of aromatic hydrocarbons than any other motor spirit. Therefore the addition of Benzole to petrol, by increasing its power-content, inevitably increases mileage per gallon. National Benzole Mixture is a scientific blend of high-grade petrol and Benzole distilled from British coal. It is the same price as petrol and its *real* cost—judged on a miles-per-gallon basis—makes it by far the most economical spirit at the disposal of the motorist to-day.

Be British in Spirit—and money in pocket!

National
Benzole Mixture

NATIONAL BENZOLE CO., LTD., Wellington House, Buckingham Gate, LONDON, S.W.1
(The distributing organisation owned and entirely controlled by the producers of British Benzole.)

The growing realisation that leaded fuel was a major contributor to global environmental harm took hold as far back as the late 1960s but moves to ban it altogether took until the 1990s to arrive in Britain. Catalysers, which by now had become a mandatory part of a car's exhaust system to reduce harmful emissions, would be damaged and rendered useless by leaded petrol, so not wanting to foot multi-million-pound bills for replacing them, the oil companies withdrew it from sale in favour of unleaded petrol. It also signalled the end of the star system.

If they were to stay on the road, older cars needed to have their engines adjusted to cope with this less-damaging petrol or be given an additive at every fill up.

Tighter restrictions imposed today on petrol and diesel to protect the environment have seen ethanol being added to fuel – a 5 per cent mix of ethanol with petrol and 7 per cent with diesel has been in use for several years. Ethanol helps maintain the octane level of petrol, and acts as an effective anti-knocking agent, while at the same time reducing the levels of carbon monoxide being emitted.

In July 2018, RAC Foundation research claimed that a new E10 petrol – containing 10 per cent ethanol – could damage the engines of 860,000 older cars on Britain's roads. At the end of 2018, E10 was not on sale in Britain but could be bought easily in France and Germany. The results of a government consultation were awaited.

150 FUELLING THE MOTORING AGE

11 A BUMPY ROAD AHEAD FOR PETROL STATIONS?

According to the statista.com website, there were 8,394 filling stations in Britain at the end of 2018 – a drop of nearly 36 per cent since 2000, when there were 13,100. That's double the percentage rate of pub closures in the UK during the same period of time. (We have actually lost 10,000 pubs since 2000, a fall of 17 per cent, according to British Beer & Pub Association figures. Happily, there are about 50,000 pubs remaining!)

The arrival of the supermarket petrol station in the mid-1990s reshaped the market – but cannot be blamed entirely for the decline of other forecourts. Besides, supermarket petrol stations also provide honest, secure employment for many thousands of people.

Close to half of the country's petrol and diesel sales were made via the big four supermarkets' sites in 2018 despite them having fewer forecourts than BP, Esso or Shell, who each had more than 1,000 sites. Of the supermarkets, Tesco had the greatest number of petrol stations with slightly more than 500 outlets and was the market leader, commanding a 16 per cent share of all sales.

As for the independents, ever tighter safety regulations, diminishing profit margins, the rising number of drive-offs – where the dishonest customer snatches a 100 per cent discount – have all proved decisive blows to the operator struggling to make ends meet.

Clockwise from top left: BP Chargemaster electric points are now being installed in Polar's monthly subscription-based public network of 7,000 units across Britain, the country's largest. BP purchased the Chargemaster company in June 2018 for £130 million. Versions of Chargemaster are sold for home use as well. (BP Chargemaster/Polar); Today's fossil fuel delivery devices are designed for practicality and fast turnaround. Three or sometimes four types of fuel are available from both sides of a pump that has built-in card machines for payment – removing the need to walk to the forecourt shop – unless you need a map book or flowers of questionable quality! (Chris Barker collection); The return of the kerbside fuelling point – EV drivers can also power up at the kerbside in parts of London through Polar's BP Chargemaster-based network. This rapid charger unit is located in Upper Richmond Road, south-west London, one of more than fifty across the capital. (BP Chargemaster/Polar)

Added to that is the hard fact that today's cars return more miles to the litre/gallon than ever before – so they don't need to pull into a petrol station as often. Modern oils, tyres and batteries, all mainstays of parts sales in times gone by, tend to last longer as well.

Nor do we appear to drive as many miles a year as in the past. According to RAC Foundation figures for 2018, the average annual private car mileage was 7,900 – double that number if you happen to have a company car – compared with 9,200 miles in 2002. In the 1980s, the average private mileage was nearer 12,000 miles a year.

This trend for people apparently staying at home – the rise of the internet must be a factor here – or being willing to rub shoulders with the unwashed hordes on public transport, can be offset when considering there are now more cars per household than ever before. Thus the nation's total mileage is spread among a greater number of cars. The foundation says there are now 31.5 million on Britain's roads – and nearly all of them need to fill up at a forecourt of some sort.

In more rural parts of Britain, simply getting to a petrol station can mean a round trip of 50 miles or more for some – compare that with having a choice of three or four petrol stations within a couple of miles of home in more urban areas. In areas of north and central Wales and the highlands and islands of Scotland, where forecourts are at their most scarce, residents' groups have bid for regeneration funding to take over or rebuild a petrol station in hope of keeping a small community together. In areas where the only pub, bank or Post Office may have been lost, the petrol station has become the last bastion of

Power, working with Shell, unveiled a hydrogen pump at the M40 services in Beaconsfield in March 2018 when it was the first of its kind to appear alongside mainstream petrol and diesel pumps. (Photographic services, Shell International)

A sign of change – drivers of electric vehicles look out for this logo when they need to power up their cars while out and about. (Photographic services, Shell International)

Opposite: Shell chose Britain as its first market in which to launch the Shell Recharge service in October 2017. Ten sites around the Home Counties, where it expected to reach the largest number of EV drivers, had opened by the end of that year. At the time 50kW devices could charge up a flat battery to 80 per cent capacity in around thirty minutes, but more powerful 150kW chargers were due to arrive in 2019 and do the same work in ten minutes. (Photographic services, Shell International)

community, some combining any or all of these roles, as well as being the place to exchange gossip with staff and customers.

It's not the brightest of pictures to paint and it's anybody's guess what petrol stations – the very name alone will be outmoded – will look or feel like in twenty years' time, let alone a hundred, but, as in all areas of business, adapting to changing needs is crucial to survival.

The ongoing battle around the globe to reduce emissions and protect our environment will see new petrol, diesel and many hybrid cars banned from sale in Britain by 2040. Several manufacturers have committed themselves to producing only electric vehicles within a few years.

Sales of electric and hybrid (petrol/electric) vehicles grow year by year – by the end of 2018 there were 150,000 EVs on Britain's roads – but that percentage is still tiny compared to the total number of cars in

London's Metropolitan Police created the world's largest fleet of hydrogen fuel cell electric police vehicles when it acquired eleven Toyota Mirai cars in March 2018. The Mirai was the world's first hydrogen-powered production car when launched in 2015. According to Toyota, the Mirai police cars cost around half of a conventional diesel squad car to run and can theoretically cover more than 300 miles on a single tank. (Toyota GB)

Britain. In early 2019, one industry pundit predicted that while there were 90,000 electric cars on Britain's roads at the time, that figure will climb to 9 million by 2030. BP reckon on 12 million by 2040.

Certainly, we are on the cusp of seeing the wholesale expansion needed in the number of electric charging points across the country to create a reliable infrastructure.

Plans took a leap forward in December 2018 when trade magazine *Forecourt Trader* revealed that Volkswagen and Tesco had agreed to create Britain's largest retail electric charging network, powered by Pod Point.

The two companies planned to roll out more than 2,400 EV charging bays across 600 Tesco stores – an average of four to each site – in a three-year programme finishing in 2021. The charging bays will be based in Tesco Extra and superstore car parks.

Customers will be able to charge their electric cars using a normal 7kW charger for free or a 'rapid' 50kW charger for a small cost in line with market rate. Many of the charging points are set to be in place by the time Volkswagen expects to launch its keenly priced electric car, the ID.3, in Britain sometime in the early part of 2020. The car is likely to have a range of close to 350 miles.

Electric power long ago shook off the image of being the sole preserve of milk floats struggling under the weight of their batteries to hit 15mph flat out with the wind behind. Without doubt, the technology is improving in leaps and bounds. Batteries can take a car further on one charge now than they could ever have done in the first years of this century.

Electric cars were running around in the early days of the internal combustion engine – and proved popular in a slower-moving city environment but lost out when their batteries proved too expensive to make. This became all too obvious once manufacturers of petrol cars slashed their costs after setting up moving assembly lines.

Carmakers have realised that being able to routinely drive at least 300 miles between charges is the psychological point at which motorists can be won over in sufficient numbers to start building a viable mass market.

Another hurdle may prove harder to clear. We are used to going to a petrol station, filling up with enough fuel for 400 or 500 miles, paying for it and being on our way again in less than ten minutes.

A journey from one end of the country to the other in an all-electric car becomes impractical if you have to make two or three stops en route to charge up again – and you can only drink so many highly calorific frothy coffees while you wait! Fully topping up the power can take hours, assuming you have been lucky enough to find somewhere that has plug-in points. However, the first 150kW charging points are gradually making their debut and offer the ability to replenish a flat battery to near maximum power in only a few minutes – matching, for the first time, how long it takes to fill up a petrol car.

Clearly, the hybrid models can reign supreme for the time being – when the electricity dries up, it flicks over to petrol for as long as you have fuel in the tank.

Electric and hybrid cars cost more than petrol or diesel models to buy but, as with all developing technologies, that gap should close over the next few years. The hope must be that batteries can be charged more quickly, both via home connections and an extensive network of public points.

In an all-electric world, consider how feasible it would be for forecourts to be stripped of their pumps and effectively become car parks providing rows of charging points. Each place would need to have several dozen spaces for charging cars to ensure a half-decent turnover rate – many sites simply do not have the room. Those that do would most likely become mini service areas, similar to those on motorways, offering fuel for humans with a variety of food and drink outlets.

At the moment, the government rakes in billions of pounds each year from the fuel duty and VAT charged in the price of a litre and so stands to see a big fall in revenue if everyone does turn to electric vehicles. It could be that duty or higher VAT will be introduced on the electricity we use to make up the shortfall.

All of this might be ultimately usurped in the longer-term future, always difficult to predict (think of those 1950s and '60s designs of cars for the year 2000!) as hydrogen technology continues its development.

A hydrogen cell-powered car puts out nothing more harmful than water through its exhaust. Filling up is carried out in a similar fashion to serving yourself with petrol and takes about the same length of time. No need to slip off for that expensive coffee while waiting for an electric point to do its work!

More importantly for HMRC, it could charge duty and VAT on hydrogen sales in the same way it does for petrol and diesel – hydrogen is measured in kilograms rather than litres. The filling station concept as we know it would be preserved or, perhaps more accurately, reinstated along with all the benefits their shops bring in being a focal point for more remote communities.

There are three models of hydrogen car on sale in this country now, each with price tags of around £60,000 and a claimed range of 370 miles using between 4–5kg of hydrogen. Having the money to buy a hydrogen car outright may seem eye-watering to most people, but this compares – as does the range – to the higher end electric cars.

In September 2015 ITM Power opened the first publicly accessible hydrogen filling station at the Advanced Manufacturing Park in Rotherham. After opening two more sites, ITM Power then worked with Shell to enable it to open its first UK hydrogen station at Cobham services on the Surrey stretch of the M25 in February 2017. Kitted out with the equipment to make its own hydrogen, there is no need to bring a bulk tanker to the site – making a useful saving on transport costs. More hydrogen sites followed, one was on the M40 services at Beaconsfield – the first in the UK to be integrated 'under the canopy' into an existing one. By the end of 2018 ITM Power had been involved in a total of eight sites spread as far apart as Swindon and the Isle of Orkney.

While the advantages of hydrogen appear irresistible to many, the technology and infrastructure to make them more viable and affordable are some years away.

Forecourts, in some places, are seeing a small turnaround in fortunes. In 2018, there were still 5,500 independents operating across Britain, although five companies owned one-third of them. Smaller stations were closing at a slower rate by 2018 too – only one or two a month, compared with one a day in 2010. There is still some mileage left for the British petrol station.

BIBLIOGRAPHY

Information included in this book has come from numerous sources and those listed are hereby acknowledged with the author's thanks:

Publications:
Shropshire Star
AA & RAC centenary histories
Autocar magazine
AA's *Drive* magazine

Websites:
www.igg.org.uk/rail
www.thinkdefence.co.uk
www.news.bbc.co.uk/onthisday
www.guardian.com/gnm-archive
www.motorwayservicesonline.co.uk
www.autoexpress.co.uk
www.transporttrust.com
www.statista.com

www.forecourttrader.co.uk
www.bpchargemaster.com
and the corporate websites of BP, Shell and Exxon Mobil.

All other material from the author's collection.

Photographs:
Every effort has been made to ensure the photographs in this book have been correctly sourced and credited. If there are any issues, please contact the author via birchingtonphotos@gmail.com who will be happy to consider amendments for future editions of this book.